Enzymes, Biocatalysis and Chemical Biology

Enzymes, Biocatalysis and Chemical Biology

Special Issue Editor
Stefano Serra

MDPI • Basel • Beijing • Wuhan • Barcelona • Belgrade • Manchester • Tokyo • Cluj • Tianjin

Special Issue Editor
Stefano Serra
Istituto di Scienze e Tecnologie
Chimiche (SCITEC)
Italy

Editorial Office
MDPI
St. Alban-Anlage 66
4052 Basel, Switzerland

This is a reprint of articles from the Special Issue published online in the open access journal *Molecules* (ISSN 1420-3049) (available at: https://www.mdpi.com/journal/molecules/special_issues/Enzymes_Biocatalysis_Chemical_Biology).

For citation purposes, cite each article independently as indicated on the article page online and as indicated below:

LastName, A.A.; LastName, B.B.; LastName, C.C. Article Title. *Journal Name* **Year**, *Article Number*, Page Range.

ISBN 978-3-03936-354-4 (Pbk)
ISBN 978-3-03936-355-1 (PDF)

© 2020 by the authors. Articles in this book are Open Access and distributed under the Creative Commons Attribution (CC BY) license, which allows users to download, copy and build upon published articles, as long as the author and publisher are properly credited, which ensures maximum dissemination and a wider impact of our publications.

The book as a whole is distributed by MDPI under the terms and conditions of the Creative Commons license CC BY-NC-ND.

Contents

About the Special Issue Editor vii

Stefano Serra
Enzymes, Biocatalysis and Chemical Biology
Reprinted from: *Molecules* **2020**, *25*, 2354, doi:10.3390/molecules25102354 1

Xiaoyu Lei, Shuangshuang Gao, Xi Feng, Zhicheng Huang, Yinbing Bian, Wen Huang and Ying Liu
Effects of GGT and C-S Lyase on the Generation of Endogenous Formaldehyde in *Lentinula edodes* at Different Growth Stages
Reprinted from: *Molecules* **2019**, *24*, 4203, doi:10.3390/molecules24234203 5

Te-Sheng Chang, Chien-Min Chiang, Yu-Han Kao, Jiumn-Yih Wu, Yu-Wei Wu and Tzi-Yuan Wang
A New Triterpenoid Glucoside from a Novel Acidic Glycosylation of Ganoderic Acid A via Recombinant Glycosyltransferase of *Bacillus subtilis*
Reprinted from: *Molecules* **2019**, *24*, 3457, doi:10.3390/molecules24193457 17

Lijuan Zhu, Linhu Zhu, Ayesha Murtaza, Yan Liu, Siyu Liu, Junjie Li, Aamir Iqbal, Xiaoyun Xu, Siyi Pan and Wanfeng Hu
Ultrasonic Processing Induced Activity and Structural Changes of Polyphenol Oxidase in Orange (*Citrus sinensis* Osbeck)
Reprinted from: *Molecules* **2019**, *24*, 1922, doi:10.3390/molecules24101922 29

Xiangxian Ying, Shihua Yu, Meijuan Huang, Ran Wei, Shumin Meng, Feng Cheng, Meilan Yu, Meirong Ying, Man Zhao and Zhao Wang
Engineering the Enantioselectivity of Yeast Old Yellow Enzyme OYE2y in Asymmetric Reduction of (*E*/*Z*)-Citral to (*R*)-Citronellal
Reprinted from: *Molecules* **2019**, *24*, 1057, doi:10.3390/molecules24061057 39

Paulo Castro, Leonora Mendoza, Claudio Vásquez, Paz Cornejo Pereira, Freddy Navarro, Karin Lizama, Rocío Santander and Milena Cotoras
Antifungal Activity against *Botrytis cinerea* of 2,6-Dimethoxy-4-(phenylimino)cyclohexa-2,5-dienone Derivatives
Reprinted from: *Molecules* **2019**, *24*, 706, doi:10.3390/molecules24040706 55

Kai Song, Yejing Wang, Yu Li, Chaoxiang Ding, Rui Cai, Gang Tao, Ping Zhao, Qingyou Xia and Huawei He
A Convenient, Rapid, Sensitive, and Reliable Spectrophotometric Assay for Adenylate Kinase Activity
Reprinted from: *Molecules* **2019**, *24*, 663, doi:10.3390/molecules24040663 67

Stefano Serra and Davide De Simeis
Fungi-Mediated Biotransformation of the Isomeric Forms of the Apocarotenoids Ionone, Damascone and Theaspirane
Reprinted from: *Molecules* **2019**, *24*, 19, doi:10.3390/molecules24010019 77

Xiangxian Ying, Jie Zhang, Can Wang, Meijuan Huang, Yuting Ji, Feng Cheng, Meilan Yu, Zhao Wang and Meirong Ying
Characterization of a Carbonyl Reductase from *Rhodococcus erythropolis* WZ010 and Its Variant Y54F for Asymmetric Synthesis of (*S*)-*N*-Boc-3-Hydroxypiperidine
Reprinted from: *Molecules* **2018**, *23*, 3117, doi:10.3390/molecules23123117 95

Xinhong Liang, Wanli Zhang, Junjian Ran, Junliang Sun, Lingxia Jiao, Longfei Feng and Benguo Liu
Chemical Modification of Sweet Potato β-amylase by Mal-mPEG to Improve Its Enzymatic Characteristics
Reprinted from: *Molecules* **2018**, *23*, 2754, doi:10.3390/molecules23112754 109

Arūnas Krikštaponis and Rolandas Meškys
Biodegradation of 7-Hydroxycoumarin in *Pseudomonas mandelii* 7HK4 via *ipso*-Hydroxylation of 3-(2,4-Dihydroxyphenyl)-propionic Acid
Reprinted from: *Molecules* **2018**, *23*, 2613, doi:10.3390/molecules23102613 123

About the Special Issue Editor

Stefano Serra is a Senior Researcher of the National Research Council of Italy (CNR) in the Istituto di Scienze e Tecnologie Chimiche (SCITEC), Milano. He received his laurea from the University of Pavia in 1995, working on the synthesis of the antitumoral triterpene saponaceolide B. One year later, he moved to the Politecnico of Milano, where he carried out studies on the use of enzymes in organic synthesis and on the stereoselective preparation of flavor and fragrances. In 2000, he was awarded a PhD in industrial chemistry from the University of Milano, and he became a researcher of the National Research Council in 2001. His scientific activity has been devoted to many aspects of organic synthesis and, above all, to the enantioselective preparation of chiral compounds and to the development of new synthetic methods.

Editorial

Enzymes, Biocatalysis and Chemical Biology

Stefano Serra

Consiglio Nazionale delle Ricerche (C.N.R.), Istituto di Scienze e Tecnologie Chimiche, Via Mancinelli 7, 20131 Milano, Italy; stefano.serra@cnr.it or stefano.serra@polimi.it ; Tel.: +39-02-2399-3076

Received: 6 May 2020; Accepted: 18 May 2020; Published: 18 May 2020

Chemical transformations that take advantage of biocatalysis are of great interest to chemists. The specific activity and selectivity of the enzymes allow them to perform different chemical reactions with high regio- and stereoselectivity, and a large number of biocatalyzed industrial processes have been already established.

At the same time, we can observe the emergence of chemical biology, namely the scientific discipline spanning the fields of chemistry and biology and dealing with chemistry applied to biology.

The aim of this Special Issue is to collect studies focused on biocatalysis applied to organic synthesis, as well as research related to chemical biology. The obtained contributions dealing with biotransformations, enzymology, the stereoselective synthesis of bioactive chemical compounds, active pharmaceutical ingredients, natural products and flavours have been collected in the present Issue.

Overall, the Issue has gathered ten research articles.

Two of these papers deal with fungal metabolism and fungi-mediated biotransformations. A first paper from Liu et al. [1] investigates the mRNA and protein expression levels and the activities of γ-glutamyl transpeptidase and L-cysteine sulfoxide lyase in correlation with the endogenous formaldehyde content in the edible mushroom *Lentinula edodes* at different growth stages. Formaldehyde is classified as a human carcinogen and can be found in natural and processed foods. Therefore, this research provided a molecular basis for understanding and controlling the endogenous formaldehyde formation in shiitake mushroom.

A second paper from Serra and De Simeis [2] describes a study on the biotransformation of seven natural occurring apocarotenoids by means of eleven selected fungal species. The substrates, namely ionone (α-, β- and γ-isomers), 3,4-dehydroionone, damascone (α- and β-isomers) and theaspirane are relevant flavour and fragrances components. The observed transformations are mainly oxidation reactions that afford oxygenated products such as hydroxy- keto- or epoxy-derivatives. A very significant feature of the study concerns the prospective applicability of the fungi-mediated biotransformation of apocarotenoids for the synthesis of high value natural flavours. Since some ionone, damascone and theaspirane isomers are available in natural form and the biotransformation of a natural precursor is considered a "natural method" of synthesis, the flavours obtained by means of the fungi-mediated reactions possess natural status and could be commercialized accordingly.

Four relevant works exploit the potential of some specific enzymes in biocatalyzed reactions. A first contribution from Wang et al. [3] reports on the glucosylation of ganoderic acid A, a bioactive triterpenoid isolated from the medicinal fungus *Ganoderma lucidum*.

A new ganoderic acid A-26-O-β-glucoside was produced from the O-glucosylation with recombinant BsGT110, a glycosyltransferase isolated from *Bacillus subtilis* ATCC 6633. BsGT110 was the first glycosyltransferases identified as catalyzing the glycosylation of triterpenoid at the C-26 position. Since triterpenoid glycosides may improve the bioactivity of the triterpenoid aglycone, this study could be regarded as a new tool in natural product synthesis.

The reduction of conjugated double bonds of citral using old yellow enzyme (OYE)-mediated biotransformation was studied by Ying et al. [4]. These researchers established that a significant increase of (R)-enantioselectivity in the (E/Z)-citral reduction was achieved by saturation mutagenesis of P76 and R330 in OYE2y. Remarkably, the variants P76M/R330H, P76G/R330H and P76S/R330H exhibited full (*R*)-enantioselectivity in the reduction of (*E*)-citral or (*E/Z*)-citral.

Castro et al. [5] exploit the activity of the enzyme laccase from *Trametes versicolor* to synthetize 2,6-dimethoxy-4-(phenylimino)cyclohexa-2,5-dienone derivatives. Ten products with different substitutions in the aromatic ring were synthetized and characterized. The 3,5-dichlorinated compound showed the highest antifungal activity against the phytopathogen *Botrytis cinerea*, while the *p*-methoxylated compound had the lowest activity. In addition, the results of this research suggested that the synthesized compounds produced damage in the fungal cell wall.

Ying et al. [6] selected a reductase for the synthesis of a specific chemical intermediate. The recombinant carbonyl reductase from *Rhodococcus erythropolis* WZ010 (ReCR) demonstrated strict (*S*)-stereoselectivity. The enzyme catalyzed the irreversible reduction of N-Boc-3-piperidone to (*S*)-N-Boc-3-hydroxypiperidine, which is the key chiral intermediate in the synthesis of ibrutinib, an active pharmaceutical ingredient. This NAD(H)-specific enzyme proved to be active within broad ranges of pH and temperature and had remarkable activity in the presence of higher concentration of organic solvents.

A further four works deal with the characterization of specific enzymes and with the study of their activity. Hu et al. [7] described the role of polyphenol oxidase in the browning reaction of orange juice. The research is also finalized to determine the methods of inactivation of the same enzymes. Fluorescence spectroscopy, circular dichroism and dynamic light scattering were used to investigate the ultrasonic effect on polyphenol oxidase activity, demonstrating that this treatment causes inactivation of the enzyme.

The correlation between activity and enzyme modification was investigated by Sun et Al. [8]. The sweet potato β-amylase was modified by six types of methoxy polyethylene glycol to enhance its specific activity and thermal stability. The aims of the study were to select the optimum modifier, optimize the modification parameters and further investigate the characterization of the modified sweet potato β-amylase. The results showed that methoxy polyethylene glycol maleimide (molecular weight 5000, Mal-mPEG5000) was the optimum modifier.

Krikštaponis et al. [9] reported a study on the 7-hydroxycoumarin catabolic pathway in *Pseudomonas* sp. 7HK4 bacteria. New metabolites and genes responsible for the degradation of 3-(2,4-dihydroxyphenyl)-propionic acid have been isolated and identified. The results show that the degradation of 7-hydroxycoumarin in *Pseudomonas* sp. 7HK4 involves a distinct metabolic pathway, compared to the previously characterized coumarin catabolic routes through a unique flavin-binding ipso-hydroxylase. Thus, the study provides new insights into the degradation of hydroxycoumarins by soil microorganisms.

Finally, He et al. [10] propose a new rapid assay to measure the activity of adenylate kinase. The latter enzyme plays a fundamental role in cellular energy and nucleotide homeostasis. The study shows a spectrophotometric analysis technique to determine enzyme activity with bromothymol blue as a pH indicator.

Overall, these ten contributions provide the reader with relevant fresh insights on the use of enzymes and on the importance of the biocatalysis. Furthermore, studies regarding the chemical biology are well represented within this Special Issue.

Funding: This research received no external funding.

Conflicts of Interest: The author declares no conflict of interest.

References

1. Lei, X.; Gao, S.; Feng, X.; Huang, Z.; Bian, Y.; Huang, W.; Liu, Y. Effects of GGT and C-S lyase on the generation of endogenous formaldehyde in *Lentinula edodes* at different growth stages. *Molecules* **2019**, *24*, 4203. [CrossRef] [PubMed]
2. Serra, S.; De Simeis, D. Fungi-mediated biotransformation of the isomeric forms of the apocarotenoids ionone, damascone and theaspirane. *Molecules* **2019**, *24*, 19. [CrossRef] [PubMed]

3. Chang, T.-S.; Chiang, C.-M.; Kao, Y.-H.; Wu, J.-Y.; Wu, Y.-W.; Wang, T.-Y. A new triterpenoid glucoside from a novel acidic glycosylation of ganoderic acid A via recombinant glycosyltransferase of *Bacillus subtilis*. *Molecules* 2019, *24*, 3457. [CrossRef] [PubMed]
4. Ying, X.; Yu, S.; Huang, M.; Wei, R.; Meng, S.; Cheng, F.; Yu, M.; Ying, M.; Zhao, M.; Wang, Z. Engineering the enantioselectivity of yeast old yellow enzyme OYE2y in asymmetric reduction of (*E/Z*)-citral to (*R*)-citronellal. *Molecules* 2019, *24*, 1057. [CrossRef] [PubMed]
5. Castro, P.; Mendoza, L.; Vásquez, C.; Pereira, P.C.; Navarro, F.; Lizama, K.; Santander, R.; Cotoras, M. Antifungal activity against *Botrytis cinerea* of 2,6-dimethoxy-4-(phenylimino)cyclohexa-2,5-dienone derivatives. *Molecules* 2019, *24*, 706. [CrossRef] [PubMed]
6. Ying, X.; Zhang, J.; Wang, C.; Huang, M.; Ji, Y.; Cheng, F.; Yu, M.; Wang, Z.; Ying, M. Characterization of a carbonyl reductase from *Rhodococcus erythropolis* WZ010 and its variant Y54F for asymmetric synthesis of (*S*)-N-boc-3-hydroxypiperidine. *Molecules* 2018, *23*, 3117. [CrossRef] [PubMed]
7. Zhu, L.; Zhu, L.; Murtaza, A.; Liu, Y.; Liu, S.; Li, J.; Iqbal, A.; Xu, X.; Pan, S.; Hu, W. Ultrasonic processing induced activity and structural changes of polyphenol oxidase in orange (*Citrus sinensis* Osbeck). *Molecules* 2019, *24*, 1922. [CrossRef] [PubMed]
8. Liang, X.; Zhang, W.; Ran, J.; Sun, J.; Jiao, L.; Feng, L.; Liu, B. Chemical modification of sweet potato β-amylase by Mal-mPEG to improve its enzymatic characteristics. *Molecules* 2018, *23*, 2754. [CrossRef] [PubMed]
9. Krikštaponis, A.; Meškys, R. Biodegradation of 7-hydroxycoumarin in *Pseudomonas mandelii* 7HK4 via ipso-hydroxylation of 3-(2,4-dihydroxyphenyl)-propionic acid. *Molecules* 2018, *23*, 2613. [CrossRef] [PubMed]
10. Song, K.; Wang, Y.; Li, Y.; Ding, C.; Cai, R.; Tao, G.; Zhao, P.; Xia, Q.; He, H. A convenient, rapid, sensitive, and reliable spectrophotometric assay for adenylate kinase activity. *Molecules* 2019, *24*, 663. [CrossRef] [PubMed]

 © 2020 by the author. Licensee MDPI, Basel, Switzerland. This article is an open access article distributed under the terms and conditions of the Creative Commons Attribution (CC BY) license (http://creativecommons.org/licenses/by/4.0/).

Article

Effects of GGT and C-S Lyase on the Generation of Endogenous Formaldehyde in *Lentinula edodes* at Different Growth Stages

Xiaoyu Lei [1], Shuangshuang Gao [1], Xi Feng [2], Zhicheng Huang [1], Yinbing Bian [3], Wen Huang [1,*] and Ying Liu [1,*]

1. College of Food Science and Technology, Huazhong Agricultural University, Wuhan 430070, China; xiaoyulei1988@126.com (X.L.); 13720161459@163.com (S.G.); HuangZhiCheng1210@163.com (Z.H.)
2. Department of Nutrition, Food Science and Packaging, California State University, San Jose, CA 95192, USA; xi.feng@sjsu.edu
3. Institute of Applied Mycology, Huazhong Agricultural University, Wuhan 430070, China; bianyb.123@163.com
* Correspondence: huangwen@mail.hzau.edu.cn (W.H.); yingliu@mail.hzau.edu.cn (Y.L.); Tel.: +86-13407161906 (Y.L.)

Academic Editor: Stefano Serra
Received: 13 October 2019; Accepted: 18 November 2019; Published: 20 November 2019

Abstract: Endogenous formaldehyde is generated as a normal metabolite via bio-catalysis of γ-glutamyl transpeptidase (GGT) and L-cysteine sulfoxide lyase (C-S lyase) during the growth and development of *Lentinula edodes*. In this study, we investigated the mRNA and protein expression levels, the activities of GGT and C-S lyase, and the endogenous formaldehyde content in *L. edodes* at different growth stages. With the growth of *L. edodes*, a decrease was found in the mRNA and protein expression levels of GGT, while an increase was observed in the mRNA and protein expression levels of C-S lyase as well as the activities of GGT and C-S lyase. Our results revealed for the first time a positive relationship of formaldehyde content with the expression levels of *Csl* (encoding Lecsl) and Lecsl (C-S lyase protein of *Lentinula edodes*) as well as the enzyme activities of C-S lyase and GGT during the growth of *L. edodes*. This research provided a molecular basis for understanding and controlling the endogenous formaldehyde formation in *Lentinula edodes* in the process of growth.

Keywords: *Lentinula edodes*; endogenous formaldehyde; GGT; C-S lyase; expression levels

1. Introduction

Lentinula edodes (shiitake mushroom) is the second-most popular edible mushrooms in the world (the No. 1 is *Agaricus bisporus*), due to its high nutritional and medicinal values as well as the unique flavor [1–4]. Lenthionine (1,2,3,5,6-pentathiepane), the unique aroma of *L. edodes* [5,6], is derived from lentinic acid in a two-step enzymatic reaction [7,8]. In the reaction, the lentinic acid is catalyzed by γ-glutamyl transpeptidase (GGT) and L-cysteine sulfoxide lyase (C-S lyase) to generate the unique flavor compounds, including lenthionine [8–10]. Nevertheless, formaldehyde is also produced in this metabolic pathway (Figure 1).

Formaldehyde, a mutagen, can be found in the air, natural and processed foods, especially in frozen food and dry foods, and is classified as a human carcinogen by the International Agency for Research on Cancer (IARC) of World Health Organization [11]. The maximum daily dose reference for formaldehyde is defined as about 0.2 mg/kg body weight per day by the US Environmental Protection Agency. Even small doses of formaldehyde can cause various symptoms of physical discomfort [12]. However, the formaldehyde contents are 1–20 mg/kg in various fruits, vegetables, meat and fish products [13]. In recent years, the amount of formaldehyde in *L. edodes* has raised the public concerns about food safety. For instance, high levels of formaldehyde (100–300 mg/kg) have been detected in shiitake mushroom samples produced in UK and Chinese [14]. Japanese researchers have reported that formaldehyde is generated in the growth process of *L. edodes* as a normal metabolite to form its unique flavor [15].

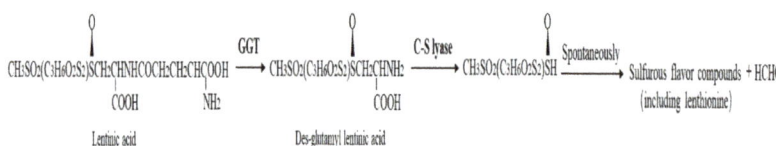

Figure 1. Proposed pathway for the generation of sulfurous flavor compounds and endogenous formaldehyde in *Lentinula edodes*.

Meanwhile, γ-glutamyl transpeptidase (GGT; EC 2.3.2.2) is an enzyme that catalyzes the transfer of the γ-glutamyl group of glutathione and related γ-glutamyl amides to water (hydrolysis) or to amino acids and peptides (transpeptidation) [16]. Cysteine sulfoxide lyase (EC 4.4.1.4) is a pyridoxal-5-phosphate (PLP) dependent enzyme and assigned to the class I family of PLP dependent enzymes [17]. In our previous work, the two enzymes were purified and characterized to play a significant biochemical role in the generation of endogenous formaldehyde in *L. edodes* [18]. Additionally, the gene of *Csl* encoding Lecsl (*L. edodes* C-S lyase) was cloned [19].

However, little is known about the relationship of GGT and C-S lyase with the production of formaldehyde in *L. edodes*. Thus, the aims of the present work were to determine the mRNA and protein expression levels and the activities of GGT and C-S lyase at different growth stages of shiitake mushrooms and to explore their correlations with endogenous formaldehyde production in *L. edodes*. This research could provide a molecular basis to understand the regulatory mechanisms of endogenous formaldehyde generation in *L. edodes* during the growth process.

2. Results and Discussion

2.1. Gene Expression of Ggtl and Csl

The expressions of *Ggtl* (encoding Leggt) and *Csl* (encoding Lecsl) during fruiting body development were analyzed by examining their transcript levels using real-time quantitative PCR. Both *Ggtl* and *Csl* were expressed at all the five stages of fruit-body development: mycelia, grey, young fruiting body, immature fruiting body and mature fruiting body, but differed in their expression patterns (Figure 2). Specifically, the transcript level of *Ggtl* decreased during the growth process, in contrast to an increase for *Csl*. Additionally, *Ggtl* showed the highest and lowest expression level in mycelia and mature fruiting body, respectively, which was just the opposite for *Csl*. There was approximately 1.5-fold difference between the highest and lowest expression levels in the two genes. Analysis of variance showed a significant difference in the *Ggtl* and *Csl* expression levels between mycelia and fruiting body stages, while *Ggtl* exhibited a significantly different expression in the four fruiting body stages ($p < 0.05$).

Based on our in-house transcriptome data, the expression pattern of *Ggtl* in different growth stages was basically in line with the quantitative RT-PCR results of our experiment. *Csl* was first reported as a gene involved in the generation of unique aroma of *L. edodes* [19]. In an early report, *Csl* displayed no obvious change in the expression level at 1, 2 and 3 h in the stage of mycelium or fruiting body during hot-air drying [20], which were basically consistent with our research results. Overall, there was a gradual decline of *Ggtl* expression level in different growth stages and an obvious increase of *Csl* expression level between mycelia and fruiting body stages.

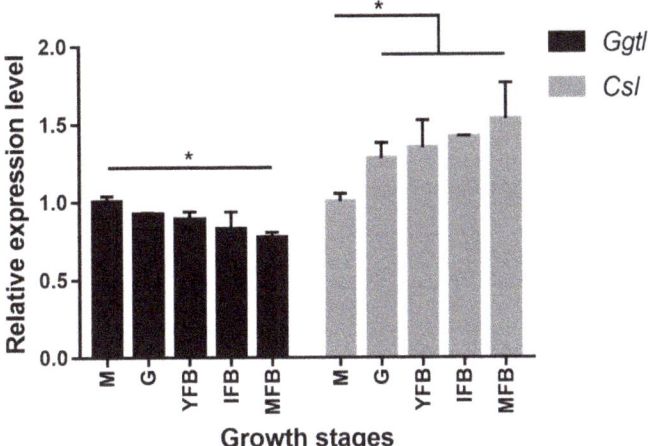

Figure 2. Expression of *Ggtl* and *Csl* during fruit-body development of *Lentinula edodes* strain W1. Relative expression during five growth stages, M, mycelia; G, grey; YFB, young fruiting body; IFB, immature fruiting body; MFB, mature fruiting body. Transcript levels of *Ggtl* (black bars) and *Csl* (gray bars) were determined by real-time quantitative RT-PCR analysis and normalized against *Actinl*. The expressions during the M stage were taken as 1. Error bars indicate standard deviation for three independent experiments. *$p < 0.05$, ANOVA tests by Duncan's indicate significant differences.

2.2. Western Blot of Leggt and Lecsl

The protein expression levels of Leggt (GGT protein of *L. edodes*) and Lecsl (C-S lyase protein of *L. edodes*) in *L. edodes* at different growth stages were determined by Western blot using β-actin as an internal reference (Figure 3A,B) [21]. Notably, Leggt exhibited three bands, which conformed to previous reports [22,23], and their gray values showed a gradual decrease in the five stages. A previous study has shown that a mature gamma-glutamyl transpeptidase consists of one polypeptide chain and can be divided into a large and a small subunit by self-catalysis at the highly conserved threonine [24]. Correspondingly, three bands were shown in our Western blot analysis of Leggt. The protein expression levels showed a decrease in Leggt while a gradual increase in Lecsl during the growth process of *L. edodes* (Figure 3C). The overall trend of protein expression levels was similar to that of mRNA expression levels. Moreover, the two proteins showed significant differences in all the five samples. The changes between the highest and lowest expression levels of Leggt and Lecsl were 2.1-fold and 1.9-fold, respectively. This is the first report on the expression levels of these two enzyme proteins in shiitake mushrooms. Our data indicated that the expression of Leggt decreased while Lecsl increased across the five growth stages of *L. edodes*.

2.3. Enzyme Activities of GGT and C-S Lyase

Figure 4A,B show the activity of GGT and C-S lyase at five different growth stages of *L. edodes*. The enzyme activities of GGT and C-S lyase were the lowest at the mycelia stage (4.7 U/g of GGT, 17.1 U/g of C-S lyase) and showed an obvious increase at the four fruiting body stages. Moreover, the GGT enzyme activity was relatively lower at the young fruiting body stage, while the C-S lyase enzyme activity showed no significant differences at the later four fruiting body stages ($p < 0.05$).

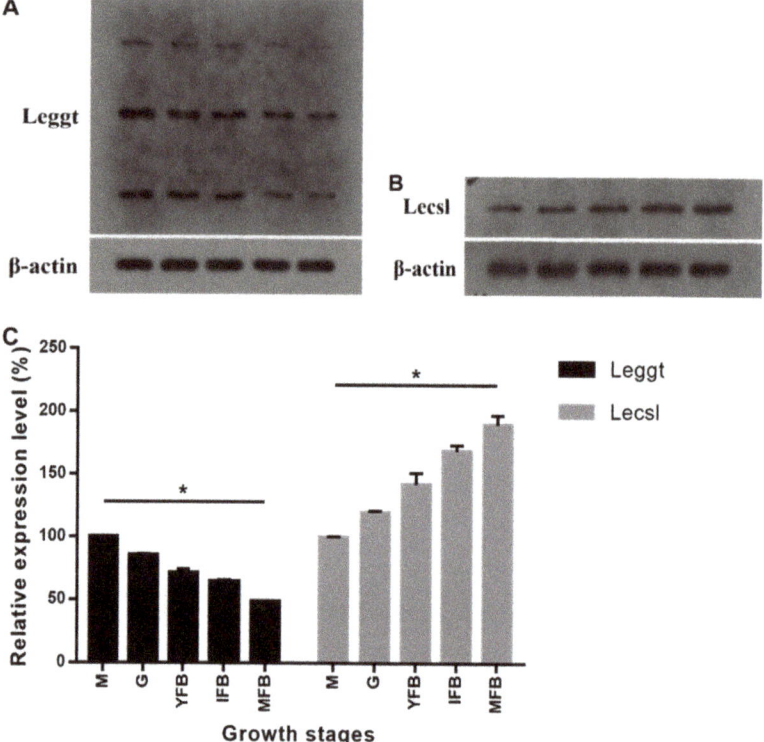

Figure 3. (**A**) Protein levels of Leggt at five stages of growth. (**B**) Protein levels of Lecsl at five stages of growth. (**C**) Relative expression of Leggt and Lecsl during five growth stages. β-actin protein was used as loading control, and the expressions during the M stage were taken as 100%. M, mycelia; G, grey; YFB, young fruiting body; IFB, immature fruiting body; MFB, mature fruiting body. Error bars indicate standard deviation for three independent experiments. *$p < 0.05$, ANOVA tests by Duncan's indicate significant differences.

Our experimental data (40.2–54.1 U/g) were consistent with the results of Huang et al., who reported that the GGT enzyme activity at the four fruiting body stages ranged from 40 to 80 U/g, with a similar difference at each stage in the fruiting body samples [25]. This is the first report about the C-S lyase enzyme activities of *L. edodes* at different growth stages. The C-S lyase enzyme activities in the present study (17.1–3575.6 U/g) were obviously lower than those determined by Xu et al. under high-temperature pre-drying (45, 55, 65 and 75 °C for 30 min) of air-dried (45 °C for 4.5 h, 60 °C for 4 h) *L. edodes* (80.17–100.54 U/mg) [26]. Liu et al. reported that the C-S lyase from *L. edodes* showed the optimum activity at 40 °C and was stable at 20–60 °C [18]. *Csl* has been demonstrated as a heat-inducible gene [27], so the activity of protein encoded by it could be improved greatly. In this study, our samples were generally collected at 25 °C, which was far below the drying treatment temperature (>45 °C), so the C-S lyase enzyme activity was lower than that treated under high temperature. Collectively, the C-S lyase enzyme activity showed no significant difference at the four fruiting body stages ($p < 0.05$).

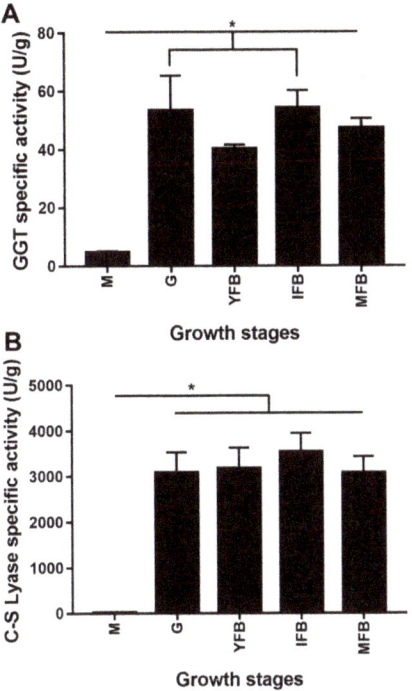

Figure 4. (**A**) The specific activity of GGT at the five growth stages. The reaction mixture containing the enzyme and the GPNA substrate was analyzed under standard conditions, and the residual activity was calculated. (**B**) The specific activity of C-S lyase at the five growth stages. The reaction mixture containing the enzyme and S-ethyl-L-cysteine sulfoxide substrate was analyzed under standard conditions, and the residual activity was calculated. M, mycelia; G, grey; YFB, young fruiting body; IFB, immature fruiting body; MFB, mature fruiting body. Error bars indicate standard deviation for three independent experiments. *$p < 0.05$, ANOVA tests by Duncan's indicate significant differences.

However, there was a marked difference between the two enzymes activities and their expression levels at the mycelia stage. GGT performs different functions in peptide transferase reaction and hydrolysis reaction under different conditions [16]. Lecsl has already been demonstrated to have one active center involved in the binding of the two substrates, S-methyl-L-cysteine sulfoxide and L-cysteine, with both cysteine sulfoxide lyase and cysteine desulfurase activities [19]. In addition, there are many factors affecting enzyme activity, such as pH, temperature, metal ions and so on. The two enzymes activities were reported to be stimulated by Na^+, K^+, Mg^{2+} and Ca^{2+} ions [18]. The environment conditions are very different between the mycelia stage and the four fruiting body stages, since the former belongs to vegetative growth and the latter belong to reproductive growth. These observations suggested that there might be a sort of regulatory mechanism that activated the two enzymes during the fruiting body stages while stayed inactive at the mycelial stage, which we failed to detect in this study.

2.4. Endogenous Formaldehyde Content in L. edodes

Figure 5 showed the content of endogenous formaldehyde in *L. edodes* at different growth phases. Compared with the mycelia stage, the endogenous formaldehyde content increased significantly ($p < 0.05$) at the four fruiting body stages, reached the maximum at the immature fruiting body stage and slightly decreased at the mature fruiting body stage. The trend of formaldehyde content at these four stages accorded with the findings of Huang et al. and Li et al. [25,28], which ranged from 13 to 89 mg/kg (dry weight) at all five stages. Mason et al. determined the formaldehyde content as 8–24 mg/kg in fresh shiitake mushrooms [29].

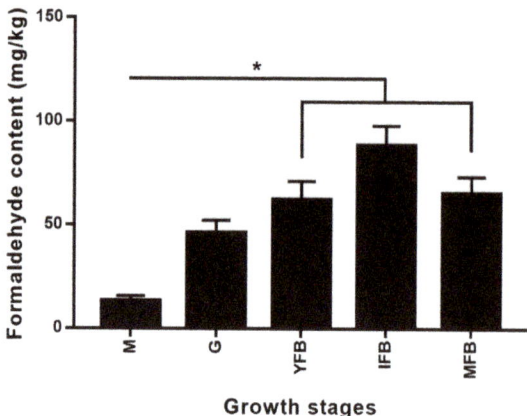

Figure 5. Endogenous formaldehyde content of *L. edodes* strain W1 at different growth stages. M, mycelia; G, grey; YFB, young fruiting body; IFB, immature fruiting body; MFB, mature fruiting body. Error bars indicate standard deviation for three independent experiments. *$p < 0.05$, ANOVA tests by Duncan's indicate significant differences.

The endogenous formaldehyde content at different growth stages showed a similar change trend to that of C-S lyase enzyme activity. The formaldehyde content of *L. edodes* during the drying has been reported to range between 150 and 400 mg/kg (dry weight) [26]. The increase of the two enzyme activities in the drying process also led to a significant increase in the endogenous formaldehyde content. Xu et al. indicated that GGT and C-S lyase were involved in formaldehyde formation and their activities were positively correlated with formaldehyde content [26]. Although the activities of the GGT and C-S Lyase were higher in IFB than in the other four stages, the differences were not significant. The endogenous formaldehyde was found to be produced from oxidative decomposition of the folate backbone and creates a benign 1C unit that can sustain essential metabolism in human cells [30]. Additionally, *L. edodes* also contains folic acid. Therefore, whether there are other enzymes and metabolic pathways involved in the generation of endogenous formaldehyde in *L. edodes* needs to be further studied.

2.5. Correlation Analysis

The effects of GGT and C-S lyase on the generation of endogenous formaldehyde in *Lentinula edodes* at different growth stages were intuitively determined by correlation analysis (Table 1). The formaldehyde content of *L. edodes* showed a positive and significant ($p < 0.01$) correlation (R) with the expression level of *Csl* and *Lecsl* and the activity of C-S lyase and GGT (0.746, 0.805, 0.867 and 0.768, respectively), while a negative relationship with the expression level of *Ggtl* and *Leggt* (−0.699 and −0.787; $p < 0.01$).

Table 1. Correlations (R) of the formaldehyde contents in *L. edodes* with *Ggtl* expression levels, *Csl* expression levels, Leggt expression levels, Lecsl expression levels, GGT enzyme activities and C-S lyase enzyme activities at different growth stages.

Properties	Formaldehyde Content
Ggtl expression levels (mRNA)	−0.699 **
Csl expression levels (mRNA)	0.746 **
Leggt expression levels (Protein)	−0.787 **
Lecsl expression levels (Protein)	0.805 **
GGT enzyme activities	0.768 **
C-S lyase enzyme activities	0.867 **

** significant at 0.01 level.

Japanese researchers pointed out that the formaldehyde content of *L. edodes* was stable during the growth process. However, the formaldehyde content after the drying process showed 3–4-fold increase. For example, in the dried shiitake mushrooms, the formaldehyde content ranged from 100 to 230 mg/kg, in contrast to 8–24 mg/kg in fresh ones [14]. Xu et al. indicated that the enzyme activities of GGT and C-S lyase were much higher under high temperature (>45 °C) than under 25 °C. These results demonstrated that the activation of the two key enzymes promoted reactions, leading to the production of a large amount of formaldehyde in *L. edodes* [26], which was well supported by our results in this study. This is the first report to show that the mRNA and protein expression levels of C-S lyase had significant and positive effects on the endogenous formaldehyde content of mushrooms.

Although the mRNA and protein expression levels of GGT were shown to be negatively correlated with the formaldehyde content, both GGT and C-S lyase were proved to be indispensable for the generation of endogenous formaldehyde in *L. edodes*. As previously reported, only the joint action of the two enzymes could promote the generation of endogenous formaldehyde [18], and GGT was the rate-limiting enzyme in the synthesis process of endogenous formaldehyde in *L. edodes* [29]. Our results showed that the activities of both of GGT and C-S lyase played a positive role in endogenous formaldehyde generation, implying the crucial effects of GGT in this process. GGT was also reported to be implicated in the transfer of amino acids across the cellular membrane and in metabolism of glutathione to cysteine by cleaving the glutamyl amide bond to preserve intracellular homeostasis by oxidative stress [31,32]. Besides, the transcription and function of genes are not synchronized in time and space. The presence of *Ggtl* homologous genes was also reported [33]. Moreover, compared with C-S Lyase, GGT has a much more complex structure and function. Despite the negative correlation of *Ggtl* and Leggt expression levels, we could not neglect their effects on the endogenous formaldehyde content in the mushroom. For a better control on the generation of endogenous formaldehyde in *L. edodes*, further studies should focus on the expression regulation of *Ggtl* and *Csl* at the transcription level.

Our study did not involve the influence of other potential metabolic pathways on the generation of endogenous formaldehyde, and whether other enzymes are implicated in the flavor metabolism pathways also needs to be investigated in future studies.

3. Materials and Methods

3.1. Fungal Strain and Culture Conditions

A dikaryotic strain of basidiomycete *Lentinula edodes* strain W1 (preserved in the Institute of Applied Mycology, Huazhong Agricultural University, Wuhan, China) was used in this study [34]. The *L. edodes* samples were obtained at five different stages: mycelia (used as control) and four fruiting body stages (grey, young fruiting body, immature fruiting body and mature fruiting body). Briefly, the mycelia were cultivated on 25 mL CYM liquid medium (2% glucose, 0.2% yeast extracts, 0.2% peptone, 0.1% K_2HPO_4, 0.05% $MgSO_4$ and 0.046% KH_2PO_4) in a conical flask and collected after growth of 12 days. Next, a conventional fruiting treatment was conducted as previously described [35]. The samples of grey (5–10 mm in cap diameter), young fruiting body (15–20 mm in cap diameter), immature fruiting body (with partial veil not ruptured) and mature fruiting body (with partial veil entirely ruptured) were harvested separately during fruiting treatment (Figure 6) [36]. The collected mushroom samples were immediately frozen in liquid nitrogen and stored at −80 °C for further use. All samples were collected in three biological replications.

Figure 6. Five growth stages of *L. edodes* strain W1. (**A**) Mycelia. (**B**) Grey (5–10 mm in cap diameter). (**C**) Young fruiting body (15–20 mm in cap diameter). (**D**) Immature fruiting body (with partial veil not ruptured). (**E**) Mature fruiting body (with partial veil entirely ruptured).

3.2. RNA Isolation and Real-Time Quantitative PCR

Total RNA was isolated using RNAiso plus (TaKaRa, Kusatsu, Japan) according to the manufacturer's instructions [37]. The total RNA concentration and purity were detected using a Nano Drop 2000 spectrophotometer (Thermo Scientific, Wilmington, DE, USA; 2.0 < A260/A280 < 2.2). The integrity of RNA was checked by electrophoresis on 1% agarose gel, and the three bands of 28S, 18S and 5S could be clearly observed (Supplementary Figure S1).

Then, 20 µL cDNA was synthesized from 1 µg of total RNA using the HiScript II Q RT SuperMix for qPCR (+ gDNA wiper) kit (Vazyme Biotech, Nanjing, China) according to the manufacturer's instructions. Next, the cDNA was two-fold diluted with double-distilled water and stored at −20 °C for quantitative RT-PCR analysis. Specific primers were designed for quantitative RT-PCR analysis of the tested genes, such as *Ggtl*, encoding Leggt (γ-glutamyl transpeptidase); *Csl*, encoding Lecsl (*L. edodes* C-S lyase, *L. edodes* genome Gene ID: LE01Gene02830) and β-actin gene (*Actinl*, encoding *L. edodes* β-actin, *L. edodes* genome Gene ID: LE01Gene01050; Supplementary Table S1) [33].

Quantitative RT-PCR was performed using a CFX Connect real-time PCR system (BIO-RAD). Each reaction consisted of 0.4 µL each of the forward and reverse primers (10 µM), 1 µL of two-fold diluted cDNA, 5 µL of 2 × Taq Master Mix (Vazyme Biotech, Nanjing, China) and 3.2 µL of double-distilled water. The qRT-PCR was performed at 95 °C for 3 min, followed by 40 cycles of 95 °C for 20 s, 60 °C for

30 s, 72 °C for 30 s and then maintaining at 72 °C for 10 min in a 96-well reaction plate. The specificity and identity of PCR products were verified by melting curve analysis to distinguish specific PCR products from the primer dimmer-caused nonspecific PCR. The existence of a single peak proved each PCR product was specific.

The relative expression was calculated using the $2^{-\Delta\Delta CT}$ method as previously described [38]. The expression of *Actinl* was used as an internal reference [39]. The expressions during the mycelium stage were taken as control. All PCR experiments were performed in three biological and three technical replications (the maximum difference in Ct was 0.5).

3.3. Extraction of Total Protein and Western Blot Analysis

Total protein of *L. edodes* was extracted as previously reported [40]. Briefly, 0.1 g of mycelia or fruiting body powder from each group (three replicates for each group) was mixed with 0.5 mL of extraction buffer (0.5 M Tris-HCl, 50 mM EDTA, 0.1 M NaCl and 40 mM dithiothreitol). The supernatants were collected after extraction for 10 min and centrifugation at $10,000 \times g$ for 15 min at 4 °C to remove the insoluble substance. Next, the same volume of saturated Tris-phenol was added to the supernatants, followed by the addition of five volumes of pre-cooled 0.1 M ammonium acetate in methanol to precipitate the protein. After washing with pre-cooled 80% acetone several times, the precipitated proteins were resolubilized and denatured for 10 min in 40 µL solution buffer (7 M urea, 50 mM Tris-HCl, 25 mM EDTA, 10 mM NaCl and 60 mM dithiothreitol). Finally, the pelleted proteins were diluted to 200 µL for further analysis. The concentration of the total protein was tested by the Coomassie Brilliant Blue G250 method [41], and the quality of protein was checked by 10% SDS-PAGE (Supplementary Figure S2) [42].

Western blot was used to analyze the expression of γ-glutamyl transpeptidase (Leggt, EC 2.3.2.2) and S-alkyl-L-cysteine sulfoxide lyase (Lecsl, EC 4.4.1.4) at different growth stages of *L. edodes*. After 50 µg of each protein sample was run on 10% SDS-PAGE gels (Bio-Rad Mini, Hercules, CA, USA), Western blot was performed by standard protocols using 1:200 anti-Leggt and anti-Lecsl polyclonal antibody sera. The antibodies against Leggt and Lecsl were raised by immunizing rabbits with the mixture of purified recombinant protein, which was expressed in *Escherichia coli* BL21 and purified by Ni-NTA Agarose column (Genscript, Nanjing, China) and Freund's adjuvant [43]. The specificity of polyclonal antibodies was detected by Western blot. The results showed that anti-Leggt polyclonal antibody sera had special bands at 68 kDa, 45 kDa and 23 kDa and anti-Lecsl polyclonal antibody sera had special band at 54 kDa, respectively. 1:50,000 horseradish peroxidase conjugated secondary antibody (BOSTER, Wuhan, China). Meanwhile, the β-actin antibody (BOSTER, Wuhan, China) was treated with the same protocol as an internal control [21].

3.4. Enzyme Activity Assays

GGT activity was determined by the transfer rate of γ-glutamyl from γ-glutamyl *p*-nitroanilide (GPNA) as reported by Liu et al. [18]. The mixture including 1 mL crude enzyme extract from *L. edodes*, 1 mL GPNA (3.5 mM) and 3 mL Tris-HCl (0.5 M, pH = 7.6) was incubated at 37 °C for 20 min and the reaction was stopped by adding 3 mL of 1.5 M cold (4 °C) acetic acid. Then, the amount of p-nitroaniline released was measured at 410 nm. The specific activity of GGT was defined as the amount of enzyme that released 1 µmol of p-nitroaniline from the substrate per min per g protein (U/g).

C-S lyase activity was measured as previously described with some modifications [44]. The mixture containing 0.3 mL crude enzyme extract from *L. edodes*, 0.5 mL S-ethyl-L-cysteine sulfoxide and 0.2 mL Tris-HCl (0.5 M, pH = 7.6) was incubated at 37 °C for 5 min. The reaction was terminated by adding 1 mL trichloroacetic acid (TCA, 10%). After supplementation with 1 mL 2,4-dinitrophenylhydrazine (DNPH, 0.1%, *m/v*) were added to the mixture was incubated for 5 min at 25 °C. Finally, 2.5 mL NaOH (2.5 M) was added to the mixture and incubated for 10 min at 25 °C. The absorbance of DNPH at 520 nm was measured. The specific activity of C-S lyase was expressed as units of enzyme per g of *L. edodes* protein (U/g).

3.5. Determination of Endogenous Formaldehyde Content in L. edodes

Steam distillation was used to extract formaldehyde from *L. edodes* at each growth stage. Each sample was supernatant of 4 g *L. edodes* homogenized with 100 mL Tris-HCl (0.5 M, pH = 7.6) buffer and 10 mL 10% (v/v) phosphoric acid aqueous solution in a 250 mL distillation flask. Water vapor was collected into a 150 mL flask, and then immersed in an ice-bath. The distillation process was stopped when 6070 mL of the distillate was collected and made up to 100 mL by deionized water. Formaldehyde in the distillate was derived by adding 1 mL of the distillate, 3.5 mL acetate buffer (0.1 M, pH = 4.0) and 0.5 mL DNPH (3 mg/mL) into a centrifuge tube at 25 °C for 15 min. Then the derived sample was filtered through a 0.22 μm filter for HPLC analysis. The formaldehyde derivative (formaldehyde-DNPH) of each group was separated and determined by a reverse-phase HPLC system (Waters, Milford, MA, USA). The mobile phase was composed of 0.05% acetic acid in acetonitrile and 0.05% acetic acid in water. The injection volume was 20 μL. All samples were detected at 355 nm as previously reported [18].

3.6. Data Analysis

All experimental data were presented as the mean ± standard deviation from at least three independent experiments. The ANOVA tests of statistical significance were performed by Duncan's multiple range tests using SPSS 20.0. p-values of <0.05 and <0.01 were accepted as significant and remarkable significant difference, respectively. The correlations of formaldehyde content with the expression levels of *Ggtl*, *Csl*, Leggt and Lecsl as well as GGT and C-S lyase activities were analyzed separately by Pearson correlation coefficient and trend of data using SPSS 20.0.

4. Conclusions

In this study, we reported for the first time the mRNA and protein expression levels and the activities of GGT and C-S lyase as well as their correlations with the endogenous formaldehyde content in *L. edodes* at different growth stages. The protein expression levels of Leggt and Lecsl were consistent with the mRNA expression levels of *Ggtl* and *Csl*. Additionally, the expression levels of GGT were decreased while those of C-S lyase were increased with the growth and development of *Lentinula edodes*. Furthermore, the enzyme activities and formaldehyde content were found to be the lowest in the mycelium stage. Our results demonstrated that the expression levels of *Csl* and Lecsl as well as the enzyme activities of C-S lyase and GGT were positively correlated with formaldehyde content during the development of *L. edodes*. These findings revealed the role of GGT and C-S lyase in generating endogenous formaldehyde at the molecular level. They also provided a molecular basis for regulating endogenous formaldehyde in the process of *L. edodes* growth.

Supplementary Materials: The supplementary materials are available online. Supplementary Figure S1: Agarose gel electrophoresis of RNA extracted from *L. edodes* at different growth stages; Supplementary Figure S2: SDS-PAGE of *L. edodes* total protein at different growth stages; Supplementary Table S1: Quantitative primer information of *Ggtl*, *Csl* and *Actinl*.

Author Contributions: X.L., Y.L. and W.H. conceived and designed the experiments. X.L., S.G. and Z.H. prepared the experiment materials. X.L. performed the experiments and analyzed the data. X.L., Y.L. and X.F. wrote the manuscript. Y.B., Y.L. and W.H. provided intellectual input and revised the manuscript. All authors read and approved the final manuscript.

Funding: This research was funded by the Natural Science Foundation of China (31601434), the Major Projects of Technological Innovation of Hubei Province (2017ABA148), the Innovation Center of Agricultural Science and Technology of Hubei Province (2016-620-000-001-044), the China Postdoctoral Science Foundation (2016T90701).

Conflicts of Interest: The authors declare no conflict of interest.

References

1. Philippoussis, A.; Diamantopoulou, P.; Israilides, C. Productivity of agricultural residues used for the cultivation of the medicinal fungus *Lentinula edodes*. *Int. Biodeter. Biodegr.* **2007**, *59*, 216–219. [CrossRef]

2. Bruhn, J.N.; Mihail, J.D.; Pickens, J.B. Forest farming of shiitake mushrooms: An integrated evaluation of management practices. *Bioresour. Technol.* **2009**, *100*, 6472–6480. [CrossRef] [PubMed]
3. Finimundy, T.C.; Dillon, A.J.P.; Henriques, J.A.P.; Ely, M.R. A review on general nutritional compounds and pharmacological properties of the *Lentinula edodes* mushroom. *Food Nutr. Sci.* **2014**, *5*, 1095–1105.
4. Choi, Y.; Lee, S.M.; Chun, J.; Lee, H.B.; Lee, J. Influence of heat treatment on the antioxidant activities and polyphenolic compounds of Shiitake (*Lentinus edodes*) mushroom. *Food Chem.* **2006**, *99*, 381–387. [CrossRef]
5. Chen, C.; Ho, C. High-performance liquid chromatographic determination of cyclic sulfur compounds of Shiitake mushroom (*Lentinus edodes* Sing.). *J. Chromatogr. A* **1986**, *356*, 455–459. [CrossRef]
6. Hiraide, M.; Miyazaki, Y.; Shibata, Y. The smell and odorous components of dried shiitake mushroom, *Lentinula edodes* I: Relationship between sensory evaluations and amounts of odorous components. *J. Wood Sci.* **2004**, *50*, 358–364. [CrossRef]
7. Yasumoto, K.; Iwami, K.; Mitsuda, H. A new sulfur-containing peptide from *Lentinus edodes* acting as a precursor for lenthionine. *Agric. Biol. Chem.* **1971**, *35*, 2059–2069. [CrossRef]
8. Yasumoto, K.; Iwami, K.; Mitsuda, H. Enzyme-catalized evolution of lenthionine from lentinic acid. *Agric. Biol. Chem.* **1971**, *35*, 2070–2080. [CrossRef]
9. Yamazaki, H.; Ogasawara, Y.; Sakai, C.; Yoshiki, M.; Makino, K.; Kishi, T.; Kakiuchi, Y. Formaldehyde in *Lentinus edodes* (in giapponese). *J. Food Hyg. Soc. Jpn.* **1980**, *21*, 165–170. [CrossRef]
10. Yasumoto, K.; Iwami, K.; Mitsuda, H. Enzymatic formation of shiitake aroma from nonvolatile precursor (s)-lenthionine from lentinic acid. *Mushroom Sci.* **1976**, *9*, 371–383.
11. Weng, X.; Chon, C.H.; Jiang, H.; Li, D. Rapid detection of formaldehyde concentration in food on a polydimethylsiloxane (PDMS) microfluidic chip. *Food Chem.* **2009**, *114*, 1079–1082. [CrossRef]
12. Huang, T.; Hong, L.; Yuan, X.; Yan, L.; Gang, Z. Preparation and characterization of a novel absorber for formaldehyde. *Proc. Int. Conf. Biol. Eng. Pharm. (BEP 2016)* **2016**, in press.
13. Tashkov, W. Determination of formaldehyde in foods, biological media and technological materials by headspace gas chromatography. *Chromatographia* **1996**, *43*, 625–627. [CrossRef]
14. Mason, D.; Sykes, M.; Panton, S.; Rippon, E. Determination of naturally-occurring formaldehyde in raw and cooked Shiitake mushrooms by spectrophotometry and liquid chromatography-mass spectrometry. *Food Addit. Contam.* **2004**, *21*, 1071–1082. [CrossRef] [PubMed]
15. Okada, S.; Iga, S.; Isaka, H. Studies on formaldehyde observed in edible mushroom shiitake, *Lentinus edodes* (Berk.) Sing (in giapponese). *J. Hyg. Chem.* **1972**, *18*, 353–357.
16. Tate, S.S.; Meister, A. γ-Glutamyl transpeptidase from kidney. *Methods Enzym.* **1985**, *113*, 400–419.
17. Kuettner, E.B.; Hilgenfeld, R.; Weiss, M.S. The active principle of garlic at atomic resolution. *J. Biol. Chem.* **2002**, *277*, 46402–46407. [CrossRef]
18. Liu, Y.; Yuan, Y.; Lei, X.Y.; Yang, H.; Ibrahim, S.A.; Huang, W. Purification and characterisation of two enzymes related to endogenous formaldehyde in *Lentinula edodes*. *Food Chem.* **2013**, *138*, 2174–2179. [CrossRef]
19. Liu, Y.; Lei, X.Y.; Chen, L.F.; Bian, Y.B.; Yang, H.; Ibrahim, S.A.; Huang, W. A novel cysteine desulfurase influencing organosulfur compounds in *Lentinula edodes*. *Sci. Rep.* **2015**, *5*, 10047. [CrossRef]
20. Gao, S.; Wang, G.Z.; Huang, Z.; Lei, X.; Bian, Y.; Liu, Y.; Huang, W. Selection of reference genes for qRT-PCR analysis in *Lentinula edodes* after hot-air drying. *Molecules* **2018**, *24*, 136. [CrossRef]
21. Rani, N.; Nowakowski, T.; Zhou, H.; Godshalk, S.E.; Lisi, V.; Kriegstein, A.; Kosik, K. A primate lncRNA mediates notch signaling during neuronal development by sequestering miRNA. *Neuron* **2016**, *90*, 1174–1188. [CrossRef] [PubMed]
22. Okada, T.; Suzuki, H.; Wada, K.; Kumagai, H.; Fukuyama, K. Crystal structure of the γ-glutamyltranspeptidase precursor protein from *Escherichia coli* structural changes upon autocatalytic processing and implications for the maturation mechanism. *J. Biol. Chem.* **2007**, *282*, 2433–2439. [CrossRef] [PubMed]
23. Boanca, G.; Sand, A.; Barycki, J.J. Uncoupling the enzymatic and autoprocessing activities of *helicobacter pylori* -Glutamyltranspeptidase. *J. Biol. Chem.* **2006**, *281*, 19029–19037. [CrossRef] [PubMed]
24. Martin, M.N.; Slovin, J.P. Purified gamma-glutamyl transpeptidases from tomato exhibit high affinity for glutathione and glutathione S-conjugates. *Plant Physiol.* **2000**, *122*, 1417–1426. [CrossRef]
25. Huang, J.; Luo, H.; Li, J. Gene cloning of γ-Glutamyltranspeptidase and its relationship to endogenous formaldehyde in shiitake mushroom (*Lentinus edodes*). *Adv. J. Food Sci. Technol.* **2016**, *12*, 579–587. [CrossRef]
26. Xu, L.; Fang, X.; Wu, W.; Chen, H.; Mu, H.; Gao, H. Effects of high-temperature pre-drying on the quality of air-dried shiitake mushrooms (*Lentinula edodes*). *Food Chem.* **2019**, *285*, 406–413. [CrossRef]

27. Huang, Z.; Lei, X.; Feng, X.; Gao, S.; Wang, G.; Bian, Y.; Huang, W.; Liu, Y. Identification of a heat-inducible element of cysteine desulfurase gene promoter in *Lentinula edodes*. *Molecules* **2019**, *24*, 2223. [CrossRef]
28. Li, J.; Song, J.; Huang, J.; Wu, N.; Zhang, L.; Jiang, T. Study on key enzymes of endogenous formaldehyde metabolism and it's content in shiitake mushrooms (*Lentinus Edodes*). *J Chin. Inst. Food Sci. Technol.* **2013**, *13*, 213–218.
29. Aberkane, H.; Frank, P.; Galteau, M.; Wellman, M. Acivicin induces apoptosis independently of gamma-glutamyltranspeptidase activity. *Biochem. Biophys. Res. Commun.* **2001**, *285*, 1162–1167. [CrossRef]
30. Burgos-Barragan, G.; Wit, N.; Meiser, J.; Dingler, F.A.; Pietzke, M.; Mulderrig, L.; Pontel, L.B.; Rosado, I.V.; Brewer, T.F.; Cordell, R.L.; et al. Mammals divert endogenous genotoxic formaldehyde into one-carbon metabolism. *Nature* **2017**, *548*, 549–554. [CrossRef]
31. Pompella, A.; Franzini, M.; Emdin, M.; Passino, C.; Paolicchi, A. Gamma-glutamylaransferase activity in human atherosclerotic plaques: Origin, prooxidant effects and potential roles in progression of disease. *Atheroscler. Supp.* **2007**, *8*, 95. [CrossRef]
32. Lim, J.S.; Yang, J.H.; Chun, B.Y.; Kam, S.; Jr, J.D.; Lee, D.H. Is serum gamma-glutamyltransferase inversely associated with serum antioxidants as a marker of oxidative stress? *Free Radic. Bio. Med.* **2004**, *37*, 1018–1023. [CrossRef] [PubMed]
33. Chen, L.; Gong, Y.; Cai, Y.; Liu, W.; Zhou, Y.; Xiao, Y.; Xu, Z.; Liu, Y.; Lei, X.; Wang, G. Genome sequence of the edible cultivated mushroom *Lentinula edodes* (Shiitake) reveals insights into lignocellulose degradation. *PLoS ONE* **2016**, *11*, e0160336. [CrossRef] [PubMed]
34. Kirk, P.M.; Cannon, P.F.; Minter, D.W.; Stalpers, J.A. Dictionary of the Fungi. *Mycol. Res.* **2009**, *113*, 908–910.
35. Gong, W.; Xu, R.; Xiao, Y.; Zhou, Y.; Bian, Y. Phenotypic evaluation and analysis of important agronomic traits in the hybrid and natural populations of *Lentinula edodes*. *Sci. Hortic.* **2014**, *179*, 271–276. [CrossRef]
36. Leung, G.S.W.; Zhang, M.; Xie, W.J.; Kwan, H.S. Identification by RNA fingerprinting of genes differentially expressed during the development of the basidiomycete *Lentinula edodes*. *Mol. Gen. Genet.* **2000**, *262*, 977–990. [CrossRef]
37. Wang, G.Z.; Ma, C.J.; Luo, Y.; Zhou, S.S.; Zhou, Y.; Ma, X.L.; Cai, Y.L.; Yu, J.J.; Bian, Y.B.; Gong, Y.H.; et al. Proteome and transcriptome reveal involvement of heat shock proteins and indoleacetic acid metabolism process in *Lentinula edodes* thermotolerance. *Cell. Physiol. Biochem.* **2018**, *50*, 1617–1637. [CrossRef]
38. Pfaffl, M.W. A new mathematical model for relative quantification in real-time RT-PCR. *Nucleic Acids Res.* **2001**, *29*, e45. [CrossRef]
39. Masaru, N.; Maki, K.; Hisayuki, W.; Machiko, O.; Kumiko, S.; Toshikazu, T.; Katsuhiro, K.; Toshitsugu, S. Important role of fungal intracellular laccase for melanin synthesis: Purification and characterization of an intracellular laccase from *Lentinula edodes* fruit bodies. *Microbiology* **2003**, *149*, 2455–2462.
40. Cai, Y.; Gong, Y.; Liu, W.; Hu, Y.; Chen, L.; Yan, L.; Zhou, Y.; Bian, Y. Comparative secretomic analysis of lignocellulose degradation by *Lentinula edodes* grown on microcrystalline cellulose, lignosulfonate and glucose. *J. Proteom.* **2017**, *163*, 92–101. [CrossRef]
41. Bradford, M.M. A rapid method for the quantitation of microgram quantities of protein utilizing the principle of protein-dye binding. *Anal. Biochem.* **1976**, *72*, 248–254. [CrossRef]
42. Laemmli, U.K. Cleavage of structural proteins during the assembly of the head of bacteriophage T4. *Nature* **1970**, *227*, 680–685. [CrossRef] [PubMed]
43. Wachino, J.; Shibayama, K.; Suzuki, S.; Yamane, K.; Mori, S.; Arakawa, Y. Profile of Expression of *Helicobacter pyloriγ*-Glutamyltranspeptidase. *Helicobacter* **2010**, *15*, 184–192. [CrossRef]
44. Yasumoto, K.; Iwami, K. S-Substituted l-cysteine sulfoxide lyase from shiitake mushroom. *Methods Enzym.* **1987**, *143*, 434–439.

Sample Availability: Samples of the constructs are available from the authors.

© 2019 by the authors. Licensee MDPI, Basel, Switzerland. This article is an open access article distributed under the terms and conditions of the Creative Commons Attribution (CC BY) license (http://creativecommons.org/licenses/by/4.0/).

Article

A New Triterpenoid Glucoside from a Novel Acidic Glycosylation of Ganoderic Acid A via Recombinant Glycosyltransferase of *Bacillus subtilis*

Te-Sheng Chang [1,†], Chien-Min Chiang [2,†], Yu-Han Kao [1], Jiumn-Yih Wu [3], Yu-Wei Wu [4,5] and Tzi-Yuan Wang [6,*]

1. Department of Biological Sciences and Technology, National University of Tainan, Tainan 70005, Taiwan; mozyme2001@gmail.com (T.-S.C.); aa0920281529@gmail.com (Y.-H.K.)
2. Department of Biotechnology, Chia Nan University of Pharmacy and Science, No. 60, Sec. 1, Erh-Jen Rd., Jen-Te District, Tainan 71710, Taiwan; cmchiang@mail.cnu.edu.tw
3. Department of Food Science, National Quemoy University, Kinmen County 892, Taiwan; wujy@nqu.edu.tw
4. Graduate Institute of Biomedical Informatics, College of Medical Science and Technology, Taipei Medical University, Taipei 106, Taiwan; yuwei.wu@tmu.edu.tw
5. Clinical Big Data Research Center, Taipei Medical University Hospital, Taipei 110, Taiwan
6. Biodiversity Research Center, Academia Sinica, Taipei 115, Taiwan
* Correspondence: tziyuan@gmail.com; Tel.: +886-2-27872258; Fax: +886-2-27899624
† The two authors contributed equally.

Academic Editor: Stefano Serra

Received: 31 August 2019; Accepted: 23 September 2019; Published: 24 September 2019

Abstract: Ganoderic acid A (GAA) is a bioactive triterpenoid isolated from the medicinal fungus *Ganoderma lucidum*. Our previous study showed that the *Bacillus subtilis* ATCC (American type culture collection) 6633 strain could biotransform GAA into compound (**1**), GAA-15-*O*-β-glucoside, and compound (**2**). Even though we identified two glycosyltransferases (GT) to catalyze the synthesis of GAA-15-*O*-β-glucoside, the chemical structure of compound (**2**) and its corresponding enzyme remain elusive. In the present study, we identified BsGT110, a GT from the same *B. subtilis* strain, for the biotransformation of GAA into compound (**2**) through acidic glycosylation. BsGT110 showed an optimal glycosylation activity toward GAA at pH 6 but lost most of its activity at pH 8. Through a scaled-up production, compound (**2**) was successfully isolated using preparative high-performance liquid chromatography and identified to be a new triterpenoid glucoside (GAA-26-*O*-β-glucoside) by mass and nuclear magnetic resonance spectroscopy. The results of kinetic experiments showed that the turnover number (k_{cat}) of BsGT110 toward GAA at pH 6 (k_{cat} = 11.2 min^{-1}) was 3-fold higher than that at pH 7 (k_{cat} = 3.8 min^{-1}), indicating that the glycosylation activity of BsGT110 toward GAA was more active at acidic pH 6. In short, we determined that BsGT110 is a unique GT that plays a role in the glycosylation of triterpenoid at the C-26 position under acidic conditions, but loses most of this activity under alkaline ones, suggesting that acidic solutions may enhance the catalytic activity of this and similar types of GTs toward triterpenoids.

Keywords: ganoderic acid A; glucosyltransferase; acidic; *Bacillus subtilis*; triterpenoid

---

1. Introduction

Ganoderma lucidum is a medicinal fungus that has been used to improve health and prevent certain diseases in Asia for thousands of years [1]. In modern ages, many bioactive compounds such as polysaccharides and triterpenoids [2,3] were identified and isolated from *G. lucidum*. These compounds were demonstrated to be effective for anti-cancer, anti-oxidant, anti-bacterial, anti-inflammation, and immune-regulation purposes [2,3].

Unlike triterpenoids from ginseng plants—which usually exist in the glycosidic form, called ginseng saponins—very few *Ganoderma* triterpenoid glycosides have been identified, despite the existence of many triterpenoids in *G. lucidum* [4]. The glycosidic form of triterpenoids might improve the bioactivity of the triterpenoids. For example, several ginseng saponins were found to exhibit more bioactivities involved in the central nervous system, cardiovascular system, and immune functions than ginseng triterpenoid aglycones were [5]. The glycosylation of flavonoids can also increase both water solubility and flavonoid stability [6–8]. It is, therefore, worthwhile to investigate the glycosylation of *Ganoderma* triterpenoids for potential medical and clinical purposes.

In nature, glycosylation is usually catalyzed by glycosyltransferase (GT, EC 2.4.x.y), a type of enzyme that uses a nucleotide-activated sugar donor, such as uridine diphosphate (UDP)-glucose, to transfer the sugar moiety to a sugar acceptor molecule [9]. Several GTs that catalyze the glycosylation of triterpenoids have already been discovered from plants, due to the accumulating knowledge on the metabolic pathways of triterpenoid glycosides [10]. However, plant GTs are not good candidates for the biotransformation of xenobiotics (such as *Ganoderma* triterpenoids) because plant GTs are usually involved in triterpenoid biosynthesis pathways and thus have very high substrate specificity. In contrast, GTs from bacterial sources usually have lower substrate specificity and have been demonstrated to be involved in the glycosylation of ginseng triterpenoids [11].

Among the *Ganoderma lucidum* bioactive compounds, ganoderic acid A (GAA) is one of the major triterpenoids and has been shown to prevent the proliferation of cancer cells and reduce inflammation activities [12–16]. Our previous study showed that the *Bacillus subtilis* ATCC (American type culture collection) 6633 strain can biotransform GAA into one major compound (**1**) and one minor compound (**2**) (Figure 1) [17]. In addition, two GTs—BsUGT398 and BsUGT489—were identified to catalyze the biotransformation of GAA into compound (**1**), which was later identified as GAA-15-O-β-glucoside [17]. However, the chemical structure of the compound (**2**) and its corresponding catalyzing enzyme remain elusive. In the present study, a GT enzyme that catalyzes the biotransformation of GAA to compound (**2**) was successfully identified, along with the optimal condition for producing compound (**2**) by the GT enzyme. The chemical structure of the previously-unknown compound (**2**) was also elucidated with the scaled-up production of the GT enzyme under an acidic condition.

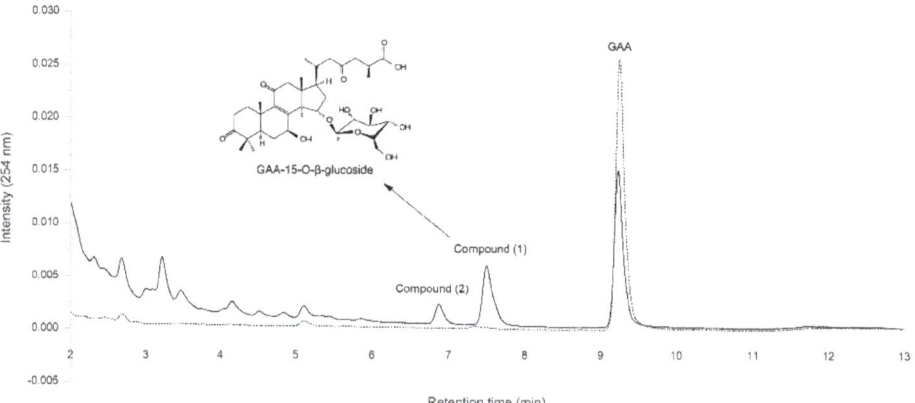

Figure 1. Biotransformation of ganoderic acid A (GAA) by *Bacillus subtilis* ATCC 6633 after 24 h of incubation (solid line). The figure was modified from Figure 1 of our previous study [17].

2. Results

2.1. Biotransformation of GAA by Recombinant BsGT110 from B. subtilis ATCC 6633

Our previous study showed that *B. subtilis* ATCC 6633 can biotransform GAA primarily into one major compound (**1**), GAA-15-O-β-glucoside, and one unknown minor compound (**2**) (Figure 1) [17]. To obtain enough unknown compound (**2**) through in vitro enzymatic biotransformation and then identify that compound's chemical structure, we strived to identify corresponding GT enzymes from the *B. subtilis* ATCC 6633 strain. In our previous work, we selected five GT genes—BsGT110, BsGT292, BsGT296, BsUGT398, and BsUGT489—and successfully overexpressed and purified them in *Escherichia coli* [17]. However, none of them were found to catalyze the biotransformation of GAA into compound (**2**) under a general GT reaction condition: 10 mM Mg^{2+}, 40 °C, and pH 8 [17]. We assayed the five recombinant BsGTs under different pH values and determined that BsGT110 produces a reasonable amount of compound (**2**) from the biotransformation of GAA under an acidic condition (pH 6), as shown in the solid line in Figure 2a. BsUGT398 and BsUGT489 produced only small amounts of compound (**2**) under the acidic condition (pH 6) (solid lines in Figure 2b,c). As expected, compound (**2**) was no longer produced from the biotransformation of GAA by any of the three GTs at pH 8 (dashed lines in Figure 2a–c). BsUGT398 and BsUGT489 produced large amounts of GAA-15-O-β-glucoside at pH 8. However, no metabolite was detected from the reactions with BsGT292 and BsGT296 at pH 6 or pH 8 (data not shown). We thus selected BsGT110 to produce compound (**2**) at pH 6 for further analysis. The amount of GAA-15-O-β-glucoside and compound (**2**) that can be catalyzed from GAA by BsGT110 at different pH values were indicated in Table 1. It is noted that the maximum amount of compound (**2**) was produced under 1 mg/mL GAA, 10 mM UDP-glucose, 15 µg/mL BsGT110, 10 mM $MgCl_2$, and 50 mM acetate buffer at pH 6.

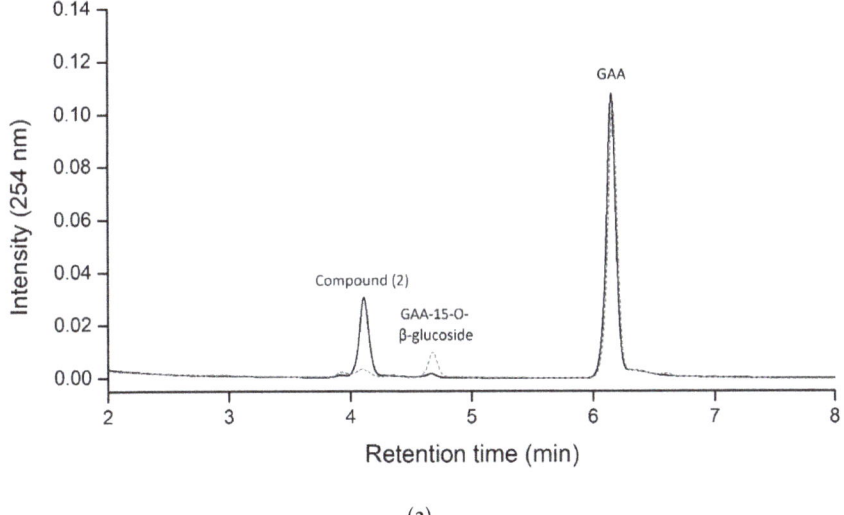

(**a**)

Figure 2. *Cont.*

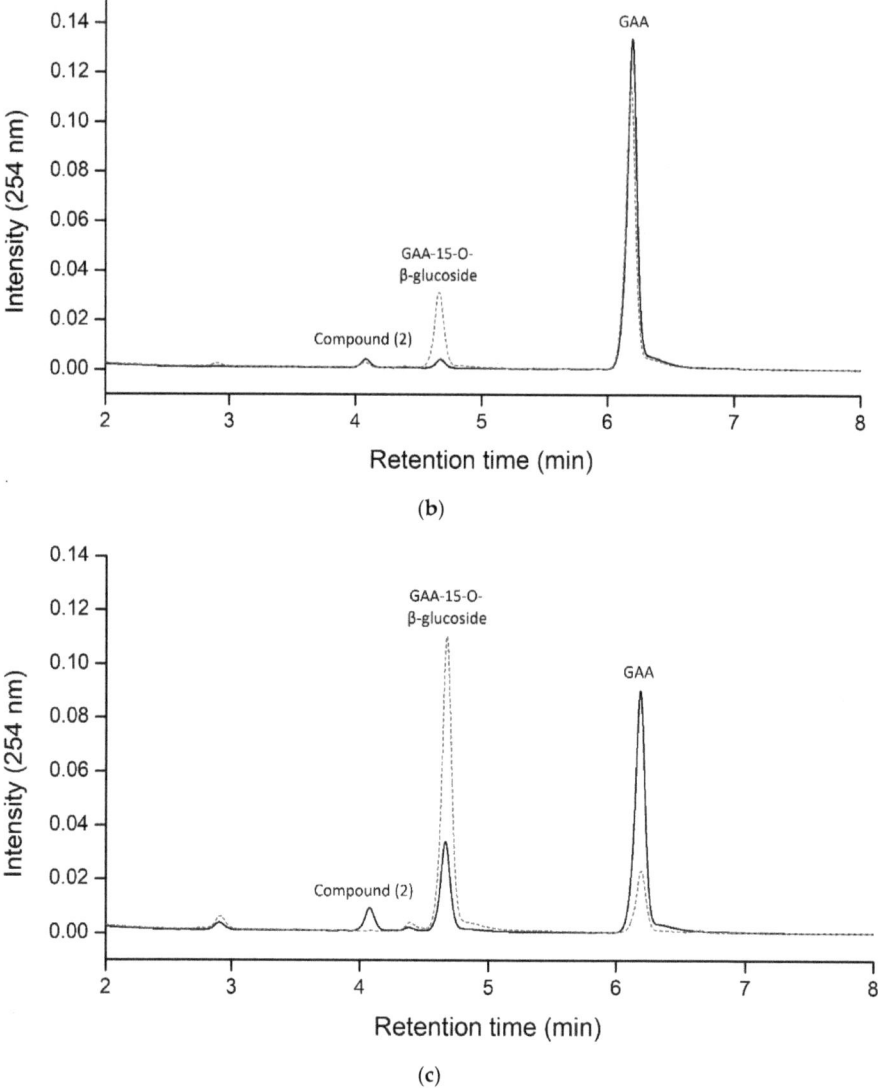

Figure 2. Ultra-performance liquid chromatography (UPLC) analysis of the biotransformation of GAA by BsGT110 (**a**), BsUGT398 (**b**), and BsUGT489 (**c**) at pH 6 (solid line) and pH 8 (dashed line). The biotransformation mixture contained 15 µg/mL purified enzyme, 1 mg/mL GAA, 10 mM uridine diphosphate (UDP)-glucose, 10 mM $MgCl_2$, and 50 mM acetate buffer at pH 6 or phosphate buffer (PB) at pH 8 and was incubated at 40 °C for 30 min. After incubation, the reaction was analyzed using UPLC. The UPLC operation procedure was described in the Materials and Methods section.

Table 1. Relative production [a] of GAA-15-O-β-glucoside and compound (2) catalyzed from GAA by BsGT110.

pH Value	Production of GAA-15-O-β-glucoside	Production of Compound (2)
5 [b]	1.13 ± 0.13	87.96 ± 4.67
6 [b]	4.31 ± 0.27	100.00 ± 7.85 [a]
6 [c]	4.85 ± 0.15	84.53 ± 9.49
7 [c]	15.26 ± 0.64	30.81 ± 0.71
8 [c]	34.02 ± 0.94	16.04 ± 1.04

[a] Relative production was normalized to the UPLC area of the peak of compound (2) in an acetate buffer of pH 6.
[b] 50 mM of acetate buffer. [c] 50 mM of PB.

To optimize the production of compound (2), a standard mixture was made of 1 mg/mL GAA, 10 mM UDP-glucose, 15 μg/mL BsGT110, and 50 mM acetate buffer at pH 6 under different temperature and metal ion conditions. After incubation, the amount of compound (2) produced was determined with UPLC (Figure 3). The results revealed that the optimal conditions for the production of compound (2) from GAA by the recombinant BsGT110 is pH 6, 40 °C, and 10 mM $MgCl_2$. The relative production of GAA-15-O-β-glucoside was less than 5% for all testing conditions.

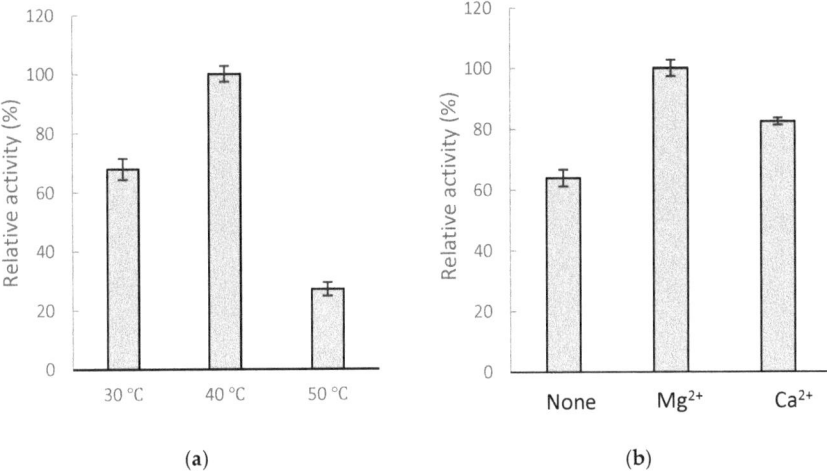

Figure 3. The production of compound (2) from GAA by BsGT110 under different temperature or metal ion conditions. The standard condition was set as 15 μg/mL purified enzyme, 1 mg/mL GAA, 10 mM $MgCl_2$, and 10 mM UDP-glucose in 50 mM acetate buffer at pH 6.0 and 40 °C. The tests were carried out by changing only the temperature (a) or metal ions (b) and maintaining all other settings. Relative activities were obtained by dividing the area summation of the UPLC reaction peak of the test condition by that of the standard condition. The data are expressed as mean ± SD, $n = 3$.

2.2. Identification of the Biotransformation Product

To resolve the chemical structure of compound (2), the biotransformation was scaled up to 25 mL, with 1 mg/mL GAA, 15 μg/mL BsGT110, 10 mM $MgCl_2$, and 10 mM UDP-glucose in 50 mM acetate buffer of pH 6 and 40 °C for a 30-min incubation. A total of 5.4 mg of compound (2) in the 25-mL reaction was purified with preparative high-performance liquid chromatography (HPLC). The chemical structure of the purified compound was then resolved using mass and nucleic magnetic resonance (NMR) spectral analyses. The molecular formula of compound (2) was established as $C_{36}H_{53}O_{12}$ by the electrospray ionization mass (ESI-MS) at m/z 679.67 $[M + H]^+$, indicating the presence of a glucose residue. The NMR spectra exhibit characteristic glucosyl signals: the anomeric carbon signal at δ_C 95.9, one CH$_2$ signal at δ$_C$ 61.8, and four CH signals at δ$_C$ 70.6, 73.9, 78.2, 79.1. The large coupling constant (8.1 Hz) of the anomeric proton H-1' (6.33 ppm) indicated the β-configuration. The cross peak of H-1' with C-26 (6.33/174.6 ppm) in the HMBC spectrum demonstrated the structure of compound (2) to be GAA-26-O-β-glucoside. The NMR spectra data are shown in Table S1 and Figures S1–S4. Figure 4 illustrated the biotransformation process of GAA by BsGT110.

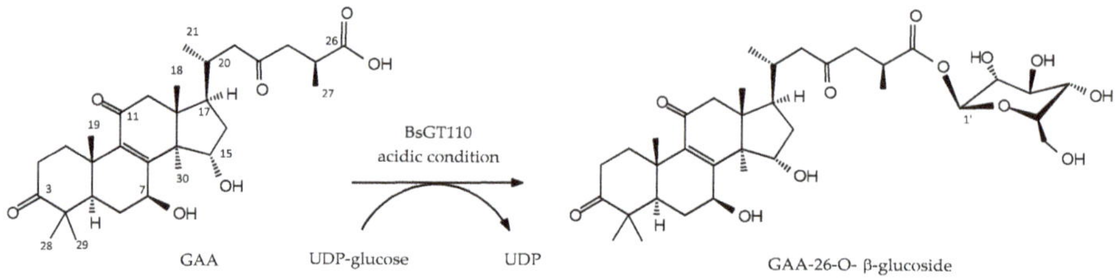

Figure 4. Biotransformation of GAA by BsGT110 in the acidic condition.

2.3. Kinetic Study of the Biotransformation of GAA by BsGT110

To study how pH affects the biotransformation activity of GAA by BsGT110, a kinetic study of the biotransformation was conducted at different concentrations of GAA, with 50 mM acetate buffer at pH 6 or PB at pH 7, 10 mM MgCl$_2$, and 10 mM UDP-glucose, at 40 °C. The results of the kinetic study were shown in Figure 5, and the calculated kinetic parameters were listed in Table 2. The results showed that the turnover number (k_{cat}) of BsGT110 toward GAA at pH 6 was 3-fold higher than that at pH 7.

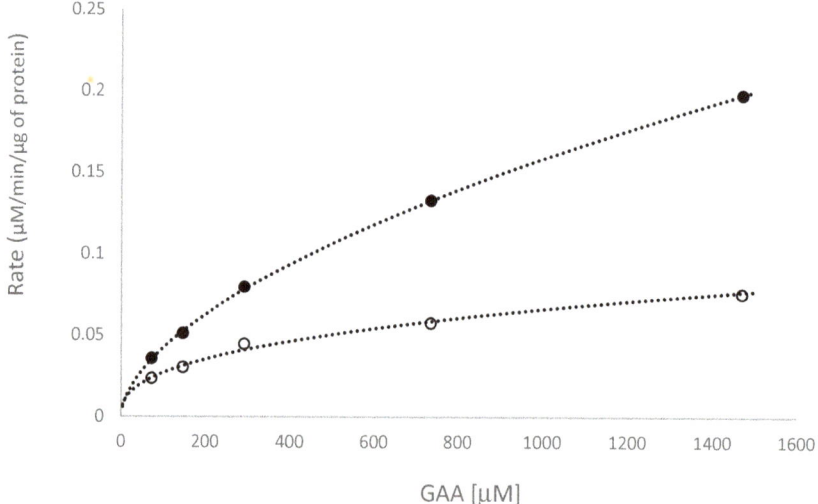

Figure 5. Kinetic study of BsGT110 at pH 6 (closed symbols) and pH 7 (open symbols). Different concentrations of GAA were mixed with 15 μg purified BsGT110 protein, 10 mM UDP-glucose, 10 mM MgCl$_2$, and 50 mM PB (pH 7.0) or acetate buffer (pH 6) in 1 mL reaction mixture and incubated at 40 °C for 20 min. During the incubation, samples from each reaction were removed and analyzed by UPLC every 2 min. The reaction rate for each concentration of GAA was obtained from the slope of the plot of the amount of product over time. The UPLC operation procedure was described in the Materials and Methods section.

Table 2. Kinetic parameters of BsGT110 toward GAA at pH 6 and pH 7.

Reaction Condition	K_m (µM)	k_{cat} (min^{-1})	k_{cat}/K_m (min^{-1}µM^{-1})
pH 6	570.6 ± 29.4	11.2 ± 0.9	0.0196 ± 0.0007
pH 7	299.4 ± 84.4	3.8 ± 0.9	0.0149 ± 0.0074

3. Discussion

According to the Carbohydrate-Active Enzymes (CAZy) database, GTs can be classified into 107 families, in which GTs that catalyze the glycosylation of small molecules, such as flavonoids and triterpenoids, are classified as GT1 [18]. Although over 500 thousands of GT have been discovered, there are only six bacterial GTs reported to catalyze glycosylation of triterpenoids, including BsYjiC from *B. subtilis* 168 [11,19–22], UGT109A1 from *B. subtilis* CTCG 63501 [23,24], BsGT1 from *B. subtilis* KCTC 1022 [25], BsUGT398 and BsUGT489 from *B. subtilis* ATCC 6633 [17], and BsGT110 from *B. subtilis* ATCC 6633 [present study]. Among them, the BsYjiC group (BsYjiC, BsUGT489, UGT109A1, and BsGT1) were highly similar, sharing more than 90% identity in their amino acid sequences [17], BsGT110 and BsUGT398, however, were not grouped with the BsYjiC group, and only had 31% and 33% identity, respectively, with the BsYjiC group (Figure 6). On the other hand, some bacterial GT1 catalyzed glycosylation of flavonoids. Thus, BsGT110 was compared with the flavonoid-catalyzing GTs. The results showed that the amino acid identity between BsGT110 and other flavonoid-catalyzing bacterial GTs was also lower than 40% (Figure S5 and Table S2). Furthermore, the evolutionary tree is shown in Figure 6 also demonstrated the dissimilarity between BsGT110 and BsUGT398, indicating that BsGT110 is a unique bacterial GT with glycosylation activity toward triterpenoids.

There are two reaction mechanisms for GT, inverting and retaining reactions, depending on the outcome of the reaction [26]. There are two stereochemical outcomes for reactions that result in the formation of a new glycosidic bond: the anomeric configuration of the product is either retained or inverted with respect to the donor substrate. The mechanistic strategy for inverting GTs involves a side chain of a residue on the active-site of GT that serves as a base catalyst that deprotonates the incoming nucleophile of the acceptor, facilitating direct displacement of the activated (substituted) phosphate leaving the group of the sugar donor, UDP-glucose [26]. Up to now, all GT1s are inverting GTs and were not reported to show optimal activities in acidic conditions [26]. For example, the well-known triterpenoid-catalyzing BsGT1 [25] had optimal activity at pH 7, BsYjiC [20], BsUGT398, and BsUGT489 [17] had optimal activity at pH 8, and UGT109A1 [23] had optimal activity at pH 9–10. These triterpenoid-catalyzing GTs had a broad neutral-alkaline range in their triterpenoid glycosylation activity [17,20–25]. According to the reaction mechanism of the inverting GTs, the side-chain of a key residue in the catalytic site of the enzyme should be deprotonated to serve as a base during the reaction. Thus, it is reasonable that GT1 enzymes have optimal activities at neutral-alkaline conditions, which would favor the deprotonation of the side chain. Accordingly, we identified the glycosylation activity of the selected five BsGTs toward GAA under a standard GT reaction condition: 10 mM Mg^{2+}, 40 °C, and pH 8, and found that only BsUGT398 and BsUGT489 can catalyze C-15 glycosylation of GAA [17], but other candidates, including BsGT110, were unable to catalyze glycosylation of GAA under the standard GT reaction condition. Thus, the previous study did not observe the novel acidic glycosylation activity (C-26 glycosylation of GAA) of BsGT110 toward GAA. Hence, the BsGT110 that we identified in this work was much more capable of catalyzing GAA into pure triterpenoid glucoside (GAA-26-O-β-glucoside) under acidic conditions (pH 5–6) (Figures 2 and 4, and Table 1). In addition, the results of the kinetics study showed that the turnover number of BsGT110 toward GAA at pH 6 was 3-fold higher than that at pH 7 (Figure 5 and Table 2). Furthermore, the catalytic efficiency (k_{cat}/K_m) of BsGT110 toward GAA at pH 6 was 1.35-fold higher than that at pH 7. Taken together, our results are unique in that they indicate that BsGT110—unlike other GT1s, which are most active at regular neutral-alkaline pH is most active at a narrow, more acidic range of pH values (pH 5–6), specifically toward the C-26 position of GAA.

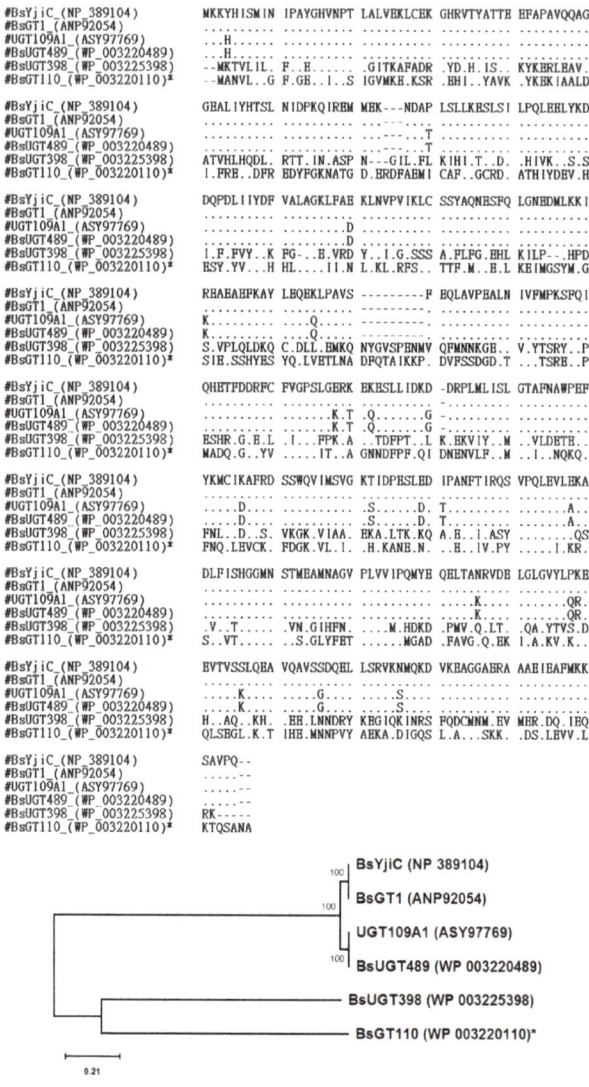

Figure 6. Aligned amino acid sequences and phylogenetic analysis using the Maximum Likelihood method. In total, 407 amino acids were aligned by Clustal W in MEGA X [27]. '.' denoted as identical amino acid, '-' denoted as indel(s). The phylogenetic tree was inferred using the Maximum Likelihood method and General Reversible Mitochondrial model [28]. The tree with the highest log likelihood (−3197.28) was shown. The percentage of trees in which the associated taxa clustered together was shown next to the branches. Initial tree(s) for the heuristic search were obtained automatically by applying Neighbor-Joining and BioNJ algorithms to a matrix of pairwise distances estimated using the JTT model, the topology with highest log-likelihood value was then selected. The tree was drawn to scale, with branch lengths measured based on the number of substitutions per site. This analysis involved six amino acid sequences. All positions with less than 95% site coverage were eliminated—i.e., less than 5% alignment gaps, missing data, and ambiguous bases were allowed at any position (partial deletion option). There were a total of 382 positions in the final dataset. Evolutionary analyses were conducted in MEGA X [27].

A few reports demonstrated that triterpenoid glycosides may improve the bioactivity of the triterpenoid aglycone [5]. Liang et al. produced an unusual ginsenoside, 3β, 12β-di-O-Glc-protopanaxadiol (PPD), from the glucosylation of PPD by UGT109A1, and showed that the ginsenoside had anti-cancer capabilities in the Lewis lung cancer xenograft mouse model [23]. Wang et al. used BsGT1 to produce 3β-O-Glc-ginsenoside F1, which inhibited melanin and tyrosinase activities [25]. Dai et al. reported the enzymatic synthesis of glycyrrhetinic acid (GA) glucosides—GA-30-O-β-glucoside and GA-3-O-β-glucoside—by BsYjiC and found that the two triterpenoid glucosides had higher water solubility and higher cytotoxicity against human liver cancer cells HepG2 and breast cancer cells MCF-7 than GA aglycone [20]. Therefore, the new GAA glucoside obtained in the present study, GAA-26-O-β-glucoside, warrants future investigation to determine whether it also has a higher bioactivity than GAA aglycone.

In summary, even though over 300 triterpenoids have been found in *G. lucidum*, very few triterpenoid glycosides have been identified [4]. Our study was the first to reveal that a single bacterium, the *Bacillus subtilis* ATCC 6633 strain, can biotransform GAA into both GAA-26-O-β-glucoside by BsGT110 in specific acidic conditions and GAA-15-O-β-glucoside by BsGT398 and BsGT489 in neutral-alkaline conditions.

4. Materials and Methods

4.1. Chemicals and Recombinant Enzymes

GAA was purchased from Baoji Herbest Bio-Tech (Xi-An, Shaanxi, China). UDP-glucose was obtained from Cayman Chemical (Ann Arbor, MI, USA). Recombinant BsGT enzymes (BsGT110, BsUGT398, BsUGT489, BsGT292, and BsGT296) were obtained from our previous studies [6,17]. The other reagents and solvents used were commercially available.

4.2. Glycosylation of GAA by Recombinant Enzymes

Glycosylation was performed in 0.1 mL reaction mixture containing 1 mg/mL GAA, 15 µg/mL the recombinant enzymes, 10 mM $MgCl_2$, and 10 mM UDP-glucose at pH 5-6 (50 mM acetate buffer) or pH 6–8 (50 mM PB). The reaction was performed at 40 °C for 30 min. Afterward, the reaction mixture was analyzed with UPLC. For optimization experiments, the above reaction mixture was incubated with 50 mM acetate buffer (pH 6) at different temperatures or with different metal ions.

For the kinetic experiments, different concentrations of GAA were mixture with 15 µg purified BsGT110 protein, 10 mM UDP-glucose, 10 mM $MgCl_2$, and 50 mM PB (pH 7.0) or acetate buffer (pH 6) in 1 mL reaction mixture and incubated at 40 °C for 20 min. During the incubation, samples from each reaction were removed every 2 min and analyzed by UPLC. The amount of GAA-26-O-β-glucoside production from the reaction was calculated from the peak area of UPLC analysis normalized with a standard curve. The rate of the reaction at each concentration of GAA was obtained from the slope of the plot of the amount of product over time. Kinetic parameters were obtained from the double-reciprocal plot of substrate GAA concentration versus the rate of reaction.

4.3. Ultra-Performance Liquid Chromatography (UPLC)

UPLC was performed with an Acquity® UPLC system (Waters, Milford, MA, USA). The stationary phase was a C18 column (Acquity UPLC BEH C18, 1.7 µm, 2.1 i.d. × 100 mm, Waters, MA, USA), and the mobile phase was 1% acetic acid in water (A) and methanol (B). The linear gradient elution condition was 0 min with 36% B to 7 min with 81% B at a flow rate of 0.2 mL/min. The detection condition was set at 254 nm.

4.4. Purification and Identification of the Glycosylated Product

Twenty-five mL of the reaction mixture (1 mg/mL GAA, 15 μg/mL BsGT110, 10 mM UDP-glucose, 10 mM MgCl$_2$, 50 mM acetate buffer at pH 6) was carried out at 40 °C for 30 min. Afterward, an equal volume of methanol was added into the reaction mixture to stop the reaction. Fifty mL of the reaction mixture with 50% methanol was applied to a preparative YL9100 HPLC system (YoungLin, Gyeonggi-do, Korea). The stationary phase was the Inertsil ODS 3 column (10 mm, 20 i.d. × 250 mm, GL Sciences, Eindhoven, The Netherlands), and the mobile phase was the same as that in the UPLC system, but with a flow rate of 15 mL/min. The detection condition was 254 nm, and the sample volume was 10 mL for each injection. The product of each run was collected, concentrated under a vacuum, and lyophilized with a freeze dryer. From the 25 mL of reaction, 5.4 mg of the product was purified. The chemical structure of the product compound was determined with mass and NMR spectral analyses. The mass spectral analysis was performed on a Finnigan LCQ Duo mass spectrometer (ThermoQuest Corp., San Jose, CA, USA) with ESI. ^{1}H- and ^{13}C-NMR, HSQC, and HMBC spectra were recorded on a Bruker AV-600 NMR spectrometer (Bruker Corp., Billerica, MA, USA) at ambient temperature. Standard pulse sequences and parameters were used for the NMR experiments, and all chemical shifts were reported in parts per million (ppm, δ).

5. Conclusions

A new GAA-26-O-β-glucoside was produced from the O-glucosylation of GAA with recombinant BsGT110 isolated from *B. subtilis* ATCC 6633 under acidic conditions (pH 6). BsGT110 was the first GT identified as catalyzing the glycosylation of triterpenoid at the C-26 position. Moreover, the optimal reaction condition of BsGT110 was at pH 6, and it lost most of this activity at pH 8, implying that such GTs might only catalyze other triterpenoid substrates under acidic conditions.

Supplementary Materials: The following are available online. Table S1. NMR spectroscopic data for compound (**2**) in pyridine-d$_5$ (600 MHz), Table S2. BsGT110 sequence comparison with candidate triterpenoid-catalyzing GTs and flavonoid-catalyzing GTs, Figure S1. The ^{1}H-NMR (600 MHz) spectrum of compound (**2**) in pyridine-d$_5$, Figure S2. The ^{13}C-NMR (150 MHz) spectrum of compound (**2**) in pyridine-d$_5$, Figure S3. The HSQC (600 MHz) spectrum of compound (**2**) in pyridine-d$_5$, Figure S4. The HMBC (600 MHz) spectrum of compound (**2**) in pyridine-d$_5$, Figure S5. Phylogenetic analysis using the Maximum Likelihood method.

Author Contributions: Conceptualization: T.-S.C. and T.-Y.W.; data curation: T.-S.C. and Y.-H.K.; methodology: Y.-H.K. and C.-M.C.; project administration: T.-S.C.; writing—original draft: T.-S.C., T.-Y.W., and C.-M.C.; writing—review and editing: T.-S.C., T.-Y.W., J.-Y.W., Y.-W.W., and C.-M.C.

Funding: This research was financially supported by grants from the Ministry of Science and Technology of Taiwan (Project No. MOST 108-2221-E-024-008-MY2).

Conflicts of Interest: The authors declare that they have no conflicts of interest.

References

1. Ahmad, M.F. *Ganoderma lucidum*: Persuasive biologically active constituents and their health endorsement. *Biomed. Pharmacother.* **2018**, *107*, 507–519. [CrossRef] [PubMed]
2. Sohretoglu, D.; Huang, S. *Ganoderma lucidum* polysaccharides as an anti-cancer agent. *Anticancer Agents Med. Chem.* **2018**, *18*, 667–674. [CrossRef] [PubMed]
3. Wu, J.W.; Zhao, W.; Zhong, J.J. Biotechnological production and application of ganoderic acids. *Appl. Microbiol. Biotechnol.* **2010**, *87*, 457–466.
4. Xia, Q.; Zhang, H.; Sun, X.; Zhao, H.; Wu, L.; Zhu, D.; Yang, G.; Shao, Y.; Zhang, X.; Mao, X.; et al. A comprehensive review of the structure elucidation and biological activity of triterpenoids from *Ganoderma* spp. *Molecules* **2014**, *19*, 17478–17535. [CrossRef] [PubMed]
5. Shi, Z.Y.; Zeng, J.Z.; Wong, A.S.T. Chemical structures and pharmacological profiles of giseng saponins. *Molecules* **2019**, *24*, 2443. [CrossRef] [PubMed]

6. Chiang, C.M.; Wang, T.Y.; Yang, S.Y.; Wu, J.Y.; Chang, T.S. Production of new isoflavone glucosides from glycosylation of 8-hydroxydaidzein by glycosyltransferase from *Bacillus subtilis* ATCC 6633. *Catalysts* **2018**, *8*, 387. [CrossRef]
7. Shimoda, K.; Hamada, H.; Hamada, H. Synthesis of xylooligosaccharides of daidzein and their anti-oxidant and anti-allergic activities. *Int. J. Mol. Sci.* **2011**, *12*, 5616–5625. [CrossRef] [PubMed]
8. Cho, H.K.; Kim, H.H.; Seo, D.H.; Jung, J.H.; Park, J.H.; Baek, N.I.; Kim, M.J.; Yoo, S.H.; Cha, J.; Kim, Y.R.; et al. Biosynthesis of catechin glycosides using recombinant amylosucrase from *Deinococcus geothermalis* DSM 11300. *Enz. Microbial Tech.* **2011**, *49*, 246–253. [CrossRef] [PubMed]
9. Hofer, B. Recent developments in the enzymatic O-glycosylation of flavonoids. *Appl. Microbiol. Biotechnol.* **2016**, *100*, 4269–4281. [CrossRef]
10. Tiwari, P.; Sangwan, R.S.; Sangwan, N.S. Plant secondary metabolism linked glycosyltransferases: An update on expanding knowledge and scopes. *Biotechnol. Adv.* **2016**, *34*, 716–739. [CrossRef]
11. Dai, L.; Li, J.; Yang, J.; Zhu, Y.; Men, Y.; Zeng, Y.; Cai, Y.; Dong, C.; Dai, Z.; Zhang, X.; et al. Use of a promiscuous glycosyltransferase from *Bacillus subtilis* 168 for the enzymatic synthesis of novel protopanaxtriol-type ginsenosides. *J. Agric. Food Chem.* **2017**, *66*, 943–949. [CrossRef] [PubMed]
12. Liang, C.; Tian, D.; Liu, Y.; Li, H.; Zhu, J.; Li, M.; Xin, M.; Xia, J. Review of the molecular mechanisms of *Ganoderma lucidum* triterpenoids: Ganoderic acids A, C2, D, F, DM, X and Y. *Eur. J. Med. Chem.* **2019**, *174*, 130–141. [CrossRef] [PubMed]
13. Jiang, J.; Grieb, B.; Thyagarajan, A.; Sliva, D. Ganoderic acids suppress growth and invasive behavior of breast cancer cells by modulating AP-1 and NF-kB signaling. *Int. J. Mol. Med.* **2008**, *21*, 577–584. [PubMed]
14. Yao, X.; Li, G.; Xu, H.; Lu, C. Inhibition of the JAK-STAT3 signaling pathway by ganoderic acid A enhances chemosensitivity of HepG2 cells to cisplatin. *Planta Med.* **2012**, *78*, 1740–1748. [CrossRef] [PubMed]
15. Wang, X.; Sun, D.; Tai, J.; Wang, L. Ganoderic acid A inhibits proliferation and invasion, and promotes apoptosis in human hepatocellular carcinoma cells. *Mol. Med. Rep.* **2017**, *16*, 3894–3900. [CrossRef] [PubMed]
16. Akihisa, T.; Nakamura, Y.; Tagata, M.; Tokuba, H.; Yasukawa, K.; Uchiyama, E.; Suzukli, T.; Kimura, Y. Anti-inflammatory and anti-tumor-promoting effects of triterpene acids and sterols from the fungus *Ganoderma lucidum*. *Chem. Biod.* **2007**, *4*, 224–231. [CrossRef] [PubMed]
17. Chang, T.S.; Wu, J.J.; Wang, T.Y.; Wu, K.Y.; Chiang, C.M. Uridine diphosphate-dependent glycosyltransferases from *Bacillus subtilis* ATCC 6633 catalyze the 15-O-glycosylation of ganoderic acid A. *Int. J. Mol. Sci.* **2018**, *19*, 3469. [CrossRef] [PubMed]
18. Cantarel, B.; Coutinho, P.M.; Rancurel, C.; Bernard, T.; Lombard, V.; Henrissat, B. The Carbohydrate-Active EnZymes database (CAZy): An expert resource for Glycogenomics. *Nucleic Acids Res.* **2009**, *37* (Suppl. 1), D233–D238. [CrossRef] [PubMed]
19. Dai, L.; Li, J.; Yang, J.; Men, Y.; Zeng, Y.; Cai, Y.; Sun, Y. Enzymatic synthesis of novel glycyrrhizic acid glucosides using a promiscuous *Bacillus* glycosyltransferase. *Catalysts* **2018**, *8*, 615. [CrossRef]
20. Dai, L.; Li, J.; Yao, P.; Zhu, Y.; Men, Y.; Zeng, Y.; Yang, J.; Sun, Y. Exploiting the aglycon promiscuity of glycosyltransferase Bs-YjiC from *Bacillus subtilis* and its application in synthesis of glycosides. *J. Biotechnol.* **2017**, *248*, 69–76. [CrossRef]
21. Li, K.; Feng, J.; Kuang, Y.; Song, W.; Zhang, M.; Ji, S.; Qiao, X.; Ye, M. Enzymatic synthesis of bufadienolide O-glycosides as potent antitumor agents using a microbial glycosyltransferase. *Adv. Synth. Catal.* **2017**, *359*, 3765–3772. [CrossRef]
22. Chen, K.; He, J.; Hu, Z.; Song, W.; Yu, L.; Li, K.; Qiao, X.; Ye, M. Enzymatic glycosylation of oleanane-type triterpenoids. *J. Asia Nat. Prod. Res.* **2018**, *20*, 615–623. [CrossRef] [PubMed]
23. Liang, H.; Hu, Z.; Zhang, T.; Gong, T.; Chen, J.; Zhu, P.; Li, Y.; Yang, J. Production of a bioactive unnatural ginsenoside by metabolically engineered yeasts based on a new UDP-glycosyltransferase from *Bacillus subtilis*. *Metab. Eng.* **2017**, *44*, 60–69. [CrossRef] [PubMed]
24. Zhang, T.T.; Gong, T.; Hu, Z.F.; Gu, A.D.; Yang, J.L.; Zhu, P. Enzymatic synthesis of unnatural ginsenosides using a promiscuous UDP-glucosyltransferase from *Bacillus subtilis*. *Molecules* **2018**, *23*, 2797. [CrossRef] [PubMed]
25. Wang, D.D.; Jin, Y.; Wang, C.; Kim, Y.J.; Perez, J.E.J.; Baek, N.I.; Mathiyalagan, R.; Markus, J.; Yang, D.C. Rare ginsenoside Ia synthesized from F1 by cloning and overexpression of the UDP-glycosyltransferase gene from *Bacillus subtilis*: Synthesis, characterization, and in vitro melanogenesis inhibition activity in BL6B16 cells. *J. Gingeng. Res.* **2018**, *42*, 42–49. [CrossRef] [PubMed]

26. Lairson, L.L.; Henrissat, B.; Davies, G.J.; Withers, S.G. Glycosyltransferases: Structures, functions, and mechanisms. *Annu. Rev. Biochem.* **2008**, *77*, 521–555. [CrossRef] [PubMed]
27. Kumar, S.; Stecher, G.; Li, M.; Knyaz, C.; Tamura, K. MEGA X: Molecular Evolutionary Genetics Analysis across Computing Platforms. *Mol. Biol. Evol.* **2018**, *35*, 1547–1549. [CrossRef] [PubMed]
28. Adachi, J.; Hasegawa, M. Model of amino acid substitution in proteins encoded by mitochondrial DNA. *J. Mol. Evol.* **1996**, *42*, 459–468. [CrossRef]

Sample Availability: 1 mg of GAA-26-O-β-glucoside for each request is available from the authors.

© 2019 by the authors. Licensee MDPI, Basel, Switzerland. This article is an open access article distributed under the terms and conditions of the Creative Commons Attribution (CC BY) license (http://creativecommons.org/licenses/by/4.0/).

Article

Ultrasonic Processing Induced Activity and Structural Changes of Polyphenol Oxidase in Orange (*Citrus sinensis* Osbeck)

Lijuan Zhu [1,2], Linhu Zhu [1,2], Ayesha Murtaza [1,2], Yan Liu [1,2], Siyu Liu [3], Junjie Li [1,2], Aamir Iqbal [1,2], Xiaoyun Xu [1,2], Siyi Pan [1,2] and Wanfeng Hu [1,2,*]

1. College of Food Science and Technology, Huazhong Agricultural University, No.1, Shizishan Street, Hongshan District, Wuhan 430070, China; lijuanzhu2019@hotmail.com (L.Z.); orion_zhu@webmail.hzau.edu.cn (L.Z.); ayeshamurtaza@webmail.hzau.edu.cn (A.M.); y20170022@mail.ecust.edu.cn (Y.L.); junjieli2019@hotmail.com (J.L.); aamirraoiqbal@webmail.hzau.edu.cn (A.I.); xiaoyunxu88@gmail.com (X.X.); drpansiyi.hzau.edu@outlook.com (S.P.)
2. Key Laboratory of Environment Correlative Dietology, Huazhong Agricultural University, Ministry of Education, Wuhan 430070, China
3. Key Laboratory of Structural Biology, School of Chemical Biology & Biotechnology, Peking University Shenzhen Graduate School, Shenzhen 518055, China; siyuliu@pku.edu.cn
* Correspondence: wanfenghu@mail.hzau.edu.cn; Tel.: +86-15071368563

Academic Editor: Stefano Serra
Received: 11 April 2019; Accepted: 14 May 2019; Published: 18 May 2019

Abstract: Apart from non-enzymatic browning, polyphenol oxidase (PPO) also plays a role in the browning reaction of orange (*Citrus sinensis* Osbeck) juice, and needs to be inactivated during the processing. In this study, the protein with high PPO activity was purified from orange (*Citrus sinensis* Osbeck) and inactivated by ultrasonic processing. Fluorescence spectroscopy, circular dichroism (CD) and Dynamic light scattering (DLS) were used to investigate the ultrasonic effect on PPO activity and structural changes on purified PPO. DLS analysis illustrated that ultrasonic processing leads to initial dissociation and final aggregation of the protein. Fluorescence spectroscopy analysis showed the decrease in fluorescence intensity leading to the exposure of Trp residues to the polar environment, thereby causing the disruption of the tertiary structure after ultrasonic processing. Loss of α-helix conformation leading to the reorganization of secondary structure was triggered after the ultrasonic processing, according to CD analysis. Ultrasonic processing could induce aggregation and modification in the tertiary and secondary structure of a protein containing high PPO activity in orange (*Citrus sinensis* Osbeck), thereby causing inactivation of the enzyme.

Keywords: browning reaction; polyphenol oxidase; ultrasonic processing; structural changes; aggregation

1. Introduction

The browning of citrus fruits during storage and juice processing often leads to undesirable flavor and nutritional loss in the final products. The reason for this browning is usually attributed to non-enzymatic browning caused by ascorbic acid degradation [1]. The enzymatic browning, which is catalyzed by polyphenol oxidase (PPO), is often ignored with regard to browning reaction in citrus products. This may be due to the presence of a high level of ascorbic acid in citrus fruits, which could reduce the colored quinone to colorless phenol, and therefore prevent the browning reaction from happening [2]. However, in long-term storage, citrus products tend to brown gradually when ascorbic acid is oxidized during storage, and the existing PPO may play its role in and be responsible for the final browning [3,4].

In previous studies, protein with high PPO activity was found in *Satsuma mandarine* juice [3,5]. In this study, PPO with high activity was also found in *Citrus sinensis* Osbeck. These enzymes accumulate in citrus peel, could easily mix with citrus juice during processing and thus catalyze the phenols to quinones, which further polymerize to generate the melanin pigments [6,7]. These colored compounds negatively affect the nutritional and organoleptic qualities, and consequently lower the marketability of citrus products [3].

Conventional methods such as thermal treatments and chemical reagents are mostly used to inhibit the browning [7]. However, thermal processing could even activate the enzyme [3,8], and may also cause loss of quality. Chemical reagents may bring safety problems to the products. Ultrasonic processing is an innovative, non-thermal technology which could retain the quality of food products at mild conditions [9]. Low-frequency, high-intensity ultrasonic processing may effectively enhance the shelf life of the juice product with minimal damage to its quality [10]. The effects of ultrasonic processing on food processing include cavitation bubbles, vibration on shear strength, and temporary generation of spots of extreme physical phenomena, as well as generation of free radicals through sonolysis of water [11]. The ultrasonic energy in liquid causes the formation of cavitation bubbles due to changes in pressure. The collision of these bubbles leads to an increase in temperature (5000 K) and pressure (1000 atm), which may generate turbulence and extreme shear force in the cavitation zone [10].

Numerous studies were conducted to investigate the effect of ultrasonic processing on enzyme inactivation during fruit and vegetable processing [7]. Up to now, literature about PPO inactivation in orange (*Citrus sinensis* Osbeck) juice has rarely been covered. The aim of the current study was to explore the effect of high-intensity ultrasonic processing on the inactivation of PPO and the structural changes of the enzyme through circular dichroism, fluorescence spectral analysis and dynamic light scattering analysis. These structural analyses may explain the mechanism of enzyme inactivation in orange (*Citrus sinensis* Osbeck) juice during ultrasonic processing.

2. Results and Discussion

2.1. Purity and Molecular Weight

As shown in Figure 1, the purified protein showed only one band after staining with R-250 Coomassie Brilliant Blue by non-denaturing (native) polyacrylamide gel electrophoresis (PAGE), which demonstrated that the protein was electrophoretically pure. Elution profile of protein extraction are shown in Figure S1. The protein purification fold and yield are showed in Table S1. The gel stained with pyrogallol showed the same protein band in the same position, confirming that the protein had PPO activity. As shown in Figure S2, it can also be confirmed that the protein had PPO activity. The protein band coincided approximately with the known protein marker of 100 kDa on the SDS and native PAGE. These results showed that the PPO might be a monomer revealing one band during electrophoresis [12]. Literature demonstrated that PPO from present molecular weight varied from 30 kDa to 128 kDa [13–15].

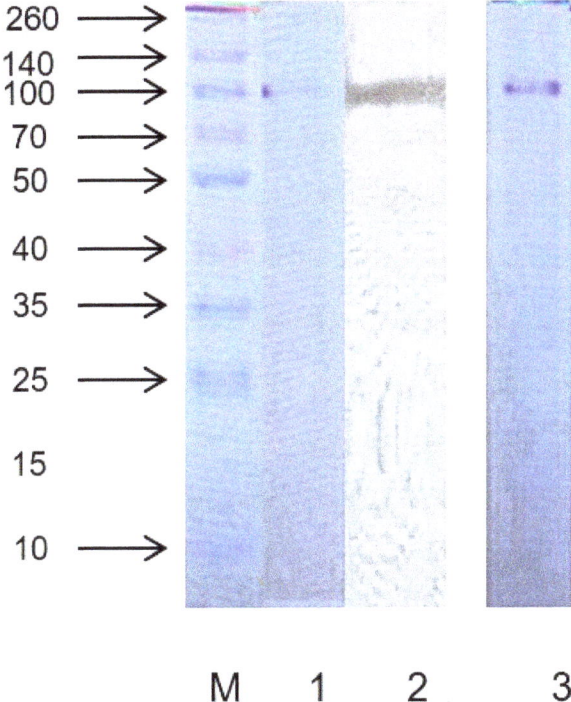

Figure 1. Electrophoresis pattern of the marker (M) and purified enzyme (1—sodium salt (SDS)-Polyacrylamide gel electrophoresis (PAGE) dyed with Coomassie Blue R-250, 2—native PAGE dyed with catechol, 3—native PAGE dyed with Coomassie Blue R-250).

2.2. Effect of Ultrasonication on PPO Activity

The PPO activities of the ultrasonicated protein solutions are presented in Figure 2A,B, respectively. As can be seen from Figure 2A, the PPO activity was very high. Ultrasonic processing at lower duration stimulated the activation of PPO. With constant power of 20 W/mL at different times (5, 10, 15, 20 and 30 min) in ultrasonic processing, the relative enzyme activity changed. The PPO activity increased with the increase of time up to 15 min and decreased with a further increase in time. As seen in Figure 2B, with constant ultrasonication for 20 min under different powers, the relative enzyme activity changed. With increasing ultrasonic power, a peak of the relative enzyme activity was observed at 20 W/mL. At above 20 W/mL, the relative enzyme activity began to reduce. Under an ultrasonic power of 40 W/mL, the enzyme activity was restrained; in particular, PPO activity was reduced by 48% under ultrasonic processing at 50 W/mL for 20 min.

These results indicate that low-intensity (20 W/mL or lower) ultrasonic processing stimulates the PPO activity, but high-intensity ultrasonic processing can inhibit the PPO activity. Previous studies demonstrated the accordant effect of ultrasonic processing on PPO activity [7,10,16]. The effect of ultrasonication on the PPO activity may be attributed to the ability of ultrasonic processing to break down the molecular aggregates. Ultrasonic processing at low intensity may dissociate protein monomers which results in the exposure of the active site to the substrate, causing the activation of the PPO enzyme [16]. However, the catalytic center was destroyed under strong ultrasonic processing and more treatment time, which might cause higher levels of denaturation and inactivation of PPO protein.

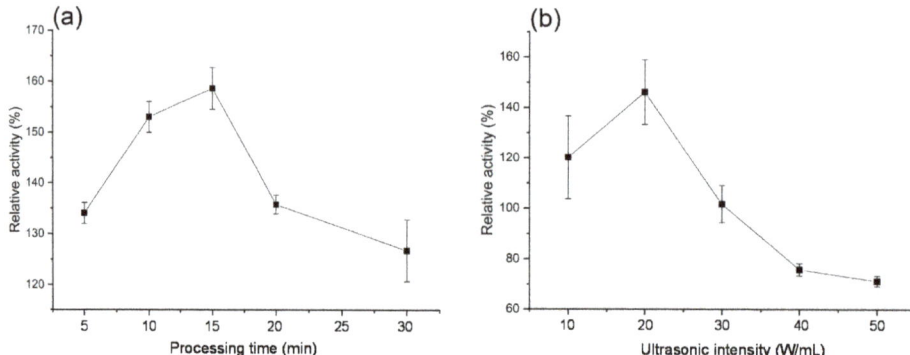

Figure 2. Relative activities of ultrasonic-processed purified PPO at 20 W/mL for 5, 10, 15, 20 and 30 min (**a**); processed for 20 min at 10, 20, 30, 40 and 50 W/mL (**b**).

2.3. Particle Size Distribution

The particle size distributions (PSD) for the untreated and ultrasonicated samples are shown in Figure 3a,b. The untreated protein has a peak value of 49.1% in number fraction at 142 nm with a relatively narrow span of 122 nm to 164 nm (42 nm), which indicates that the native protein aggregate was monodispersed in the aqueous systems [5].

The effect of ultrasonic time (5, 10, 15, 20 and 30 min at 20 W/mL) on the PSD pattern of purified protein is shown in Figure 3a. With ultrasonication for 5 min, the peak particle diameter did not change obviously, but the diameter span increased and the minimum particle diameter decreased. With increased ultrasonic time, the peak particle size and diameter span increased. The maximum peak particle size (255 nm) and diameter span (620.7 nm) were obtained by increasing ultrasonic duration to 30 min. The influence of increasing ultrasonic intensity (10 W/mL to 50 W/mL for 20 min) on PPO is shown in Figure 3b. With the intensity of 10 and 20 W/mL, the peak particle diameter did not obviously change, but the diameter span increased. When the ultrasonic intensity increased to 30 W/mL, the particles started aggregation to attain higher particle diameter. The maximum peak diameter (342 nm) and diameter span (863.7 nm) was obtained after ultrasonic processing at 50 W/mL for 20 min.

In general, after ultrasonic processing, the PSD patterns of protein exhibited remarkable changes with wider spans of larger and smaller particle diameters, suggesting the polydispersion and aggregation of protein particles [17]. The native protein aggregate was monodispersed in the aqueous systems because of hydrogen bonding, hydrophobic interaction and electrostatic interaction. A previous study found that ultrasonic processing caused the cleavage of agglomerates, hydrophobic interaction, and van der Waal's forces among protein molecules, thereby inducing the structural changes in enzyme protein [18]. In this case, the particle sizes were decreased because of the shear forces due to cavitation [18,19]. However, as the ultrasonic processing duration and intensity increased, some particles started aggregation. Similarly, Gulseren et al. [20] found that ultrasonic processing at a high duration of 40 min could cause the formation of large aggregates in bovine serum albumin protein. This aggregation might be due to the electrostatic and hydrophobic non-covalent interactions among protein particles [20]. The surface hydrophobicity decreased after prolonged sonication on reconstituted whey protein concentrate, which is a sign of protein aggregation [21]. Under the low-intensity or short-term ultrasonic processing, the enzyme exhibited dissociation because of the forces caused by cavitation. However, the aggregation increased with the increased intensity and time due to noncovalent interactions.

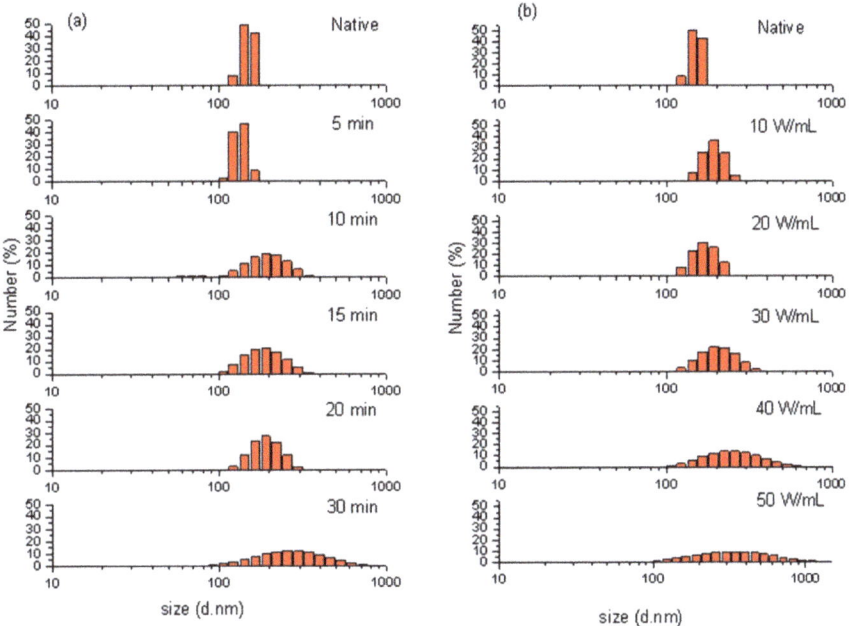

Figure 3. Particle size distributions of native and ultrasonic-processed PPO at 20 W/mL for 5, 10, 15, 20 and 30 min (**a**); processed for 20 min at 10, 20, 30, 40 and 50 W/mL (**b**).

2.4. Fluorescence Spectroscopic Analysis

The fluorescence property of the native and ultrasonicated purified enzyme was studied by fluorescence spectroscopy. It is a valuable technique to investigate the transition in the tertiary structure of proteins because the fluorescence from the tryptophan of amino acid is sensitive to the polarity of its local environment [16,22]. As shown in Figure 4, the λ_{max} of the native protein was 334 nm with an intensity of 39.74, signifying that Trp residues in enzymes were located in the nonpolar hydrophobic environment. Figure 4a shows the effect of the ultrasonic duration (5, 10, 15, 20 and 30 min at 20 W/mL) on the intrinsic fluorescence spectrum of protein. The ultrasonically treated enzyme showed a red-shift of 0.6–1.2 nm in λ_{max}, with a gradual decrease in fluorescence intensity following the increasing treatment duration. This finding indicated the exposure of Trp residues to the polar environment, thereby disrupting the tertiary structure. More exposure of fluorophores to the polar environment might cause the release and transfer of energy, which consequently leads to the quenching of fluorescence intensity [23,24].

Figure 4B illustrates the effect of increasing ultrasonic intensities (10 W/mL to 50 W/mL) on the fluorescence spectra of purified protein. The purified protein showed red-shifts of 0.2–2.0 nm in λ_{max} when treated with low ultrasonic intensities of 10, 20 and 30 W/mL. In the meantime, the fluorescence intensity gradually decreased with increasing ultrasonic intensity. The fluorophores were exposed to a more polar environment due to the polydispersity of protein aggregate after ultrasonic processing as described above, which caused the red-shift and consequently led to the quenching of fluorescence intensity. However, the extreme ultrasonic processing at 50 W/mL induced aggregation and may bury the exposed fluorophores inside the molecules, thereby decreasing the λ_{max}. PPO structural change is responsible for the activity change. The changes in fluorescence intensity and λ_{max} indicated a possible change in PPO's tertiary structure, which ultimately led to the activity reduction of PPO [25]. The result was similar to the observation that the fluorescence intensity of mushroom tyrosinase decreased in an aqueous system following mild thermal and supercritical CO_2 treatments [24].

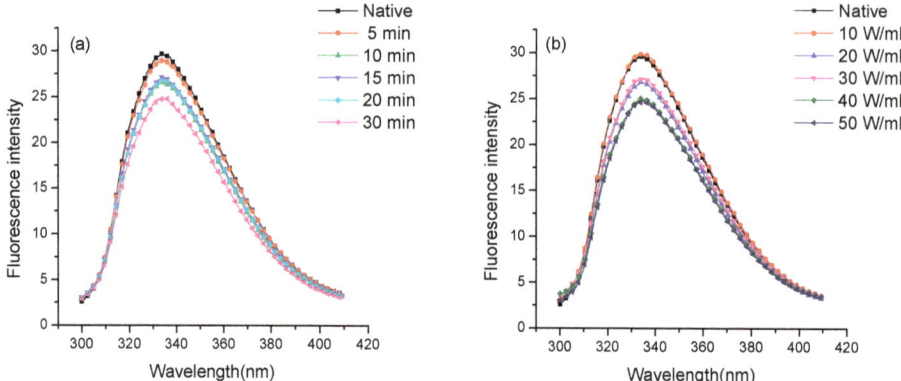

Figure 4. Fluorescence spectra of native and ultrasonic-processed PPO at 20 W/mL for 5, 10, 15, 20 and 30 min (**a**); processed for 20 min at 10, 20, 30, 40 and 50 W/mL (**b**).

2.5. Circular Dichroism Spectroscopy Analysis

The secondary structures of the native and ultrasonicated enzyme were analyzed through CD spectroscopy. The CD spectra of protein are shown in Figure 5a,b. The native protein showed a positive peak at 193 nm with two double-negative slots (208 and 222 nm), which were considered as typical α-helix conformation in the secondary structures [16,26,27]. As shown in Figure 5a,b, the negative peak (208 nm) increased with the negative peak decrease at 222 nm after ultrasonic processing. As the ultrasonic time and intensity increased, the change grew and the two double-negative slots gradually disappeared. These changes show that the ultrasonic processing triggered α-helix conformation of secondary structure loss [3,16,26–28]. The contents based on the CONTIN algorithms of the protein secondary structure of the native and ultrasonicated protein [24] are shown in Table 1. At the high ultrasonic intensity of 40 W/mL, the α-helix conformation remarkably decreased, while β-turn contents were increased. A similar study was reported by Liu et al. [10], where ultrasonic processing at high intensity caused a loss of α-helix conformation in protein structure. The decrease in α-helix was found to be correlated with the enzymatic activity of PPO molecules [29]. Ultrasonic processing induced the changes in molecular interaction, thus leading to changes in secondary structure and eventually causing the loss of PPO activity.

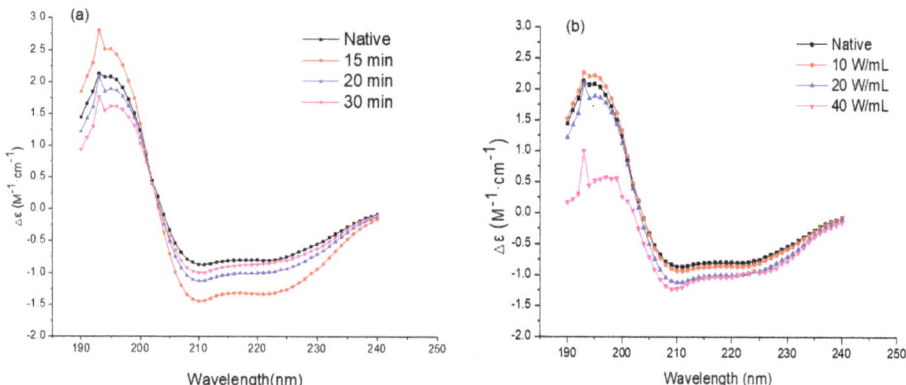

Figure 5. Circular Dichroism (CD) spectra of native and ultrasonic-processed PPO at 20 W/mL for 15, 20 and 30 min (**a**); processed for 20 min at 10, 20 and 40 W/mL (**b**).

Table 1. Secondary structure contents of native and ultrasonic-processed PPO.

Samples	α-Helix	β-Sheet	β-Turn	Random Coil
Native	76.20%	0.00%	23.60%	0.10%
10 W/mL/20 min	76.00%	0.00%	24.00%	0.00%
20 W/mL/20 min	53.30%	0.00%	29.30%	17.50%
40 W/mL/20 min	32.80%	0.00%	33.60%	33.60%
20 W/mL/15 min	54.60%	0.00%	26.50%	18.90%
20 W/mL/20 min	53.30%	0.00%	29.30%	17.50%
20 W/mL/30 min	51.80%	0.00%	31.10%	17.20%

3. Materials and Methods

3.1. Materials and Chemicals

Fresh orange (*Citrus sinensis* Osbeck) used in this study was procured from a local market (Chongqing, China). All chemicals used were of analytical grade.

3.2. Extraction and Partial Purification of Protein

Orange tissue (250 g) was homogenized in 700 mL of Tris-HCl buffer (0.5 mol/L) containing 10% polyvinylpyrrolidone (PVPP). The homogenate liquid was stored for 8 h at 4 °C and then subjected to centrifugation at 2057× g for 10 min using a centrifuge machine (Eppendorf centrifuge 5804 R, Eppendorf, Hamburg, Germany). The resultant supernatant was fractionated with 25% solid ammonium sulfate to remove impurities. The process was repeated using 90% ammonium sulfate saturation to precipitate proteins with PPO activity. The precipitate was re-dissolved in Tris-HCl (0.5 mol/L) and dialyzed against the same buffer for 34 h. The buffer was changed every hour during the whole process of dialysis. Ultrafilter was used to concentrate the crude extract of protein. Then, DEAE Sepharose Fast Flow and Sepracryl S-200 columns were used to purify the crude protein [5]. The fractions containing the highest activity of PPO were selected, concentrated and stored for further analysis.

3.3. Electrophoresis Assay

Native PAGE was carried out on preparative 12% polyacrylamide gels using the method described by Davis [30] with slight modifications. After running, the gels were stained with 0.1 mol/L Catechol (50 mL) and Coomassie Brilliant Blue R-250. The gels were analyzed for activity and estimation of molecular weight. The molecular weight and subunit of purified PPO enzyme were determined by SDS–PAGE [31].

3.4. Protein Content

Protein content was determined according to the Bradford method [32]. The protein solution was stained with Coomassie Brilliant Blue G-250 and the absorption peak was observed at 595 nm wavelength [33]. The absorption value of A595 was directly proportional to the protein concentration. The standard curve of bovine serum protein (BSA) was used as a standard protein. The concentration of sample protein was calculated according to the standard curve.

3.5. Ultrasonic Processing

An ultrasonic processor (JY92-2D, Ningbo, China) containing a titanium probe of diameter 0.636 cm was used to sonicate 10 mL of protein in 25 mL of centrifugal tubes, surrounded by ice to maintain the low temperature. All ultrasonic processing was definitely below 30 °C. The protein samples were treated at a low frequency of 20 kHz with the pulse duration of 5 s on and 5 s off setting, to investigate the effect of ultrasonic time (10, 15, 20 and 30 min at 20 W/mL) and intensity (10,

20 and 40 W/mL for 30 min) on the protein. The ultrasonic-processed samples were stored at 4 °C for further analysis.

3.6. PPO Activity Assay

The activity of the PPO enzyme was measured by determining the increasing rate of absorbance per minute at 420 nm using an Eppendorf Bio-spectrometer (Eppendorf, BioSpectrometer kinetic, Hamburg, Germany). The protein solution of 0.5 mL was mixed with 0.1 mol/L catechol (1 mL) and 0.1 mol/L Tris-HCl buffer (1 mL), then the absorbance of the resultant mixture was measured at 420 nm [5]. The relative activity (RA) of the PPO enzyme was measured according to the following equation.

$$\text{Relative PPO activity} = \frac{PPO\ activity\ of\ ultrasonic\ treated\ ppo}{PPO\ activity\ of\ native\ ppo} \times 100\% \quad (1)$$

3.7. Particle Size Distribution Analysis

Particle size determination was performed using a zetasizer, Nano-ZS device (Malvern Instruments, Malvern, Worcestershire, UK). The purified protein solutions prepared in Tris–HCl buffers (50 mmol/L) were subjected to scattered light wavelength (532 nm) and 15° laser reflection angle at a measuring temperature of 25 °C. The size measurements were taken as the mean of five readings.

3.8. CD Spectral Measurement

The purified protein solutions were subjected to CD spectra measurement with a spectropolarimeter (JASCO J-1500, Tokyo, Japan). The samples of treated protein solutions were prepared in Tris–HCl buffer (50 mmol/L) using this buffer as a blank. The secondary structure of protein solutions was determined in the range of far-ultraviolet (196–260 nm) with the scanning speed of 120 nm/min and the bandwidth of 1 nm. Data for CD spectra were presented as changes in the molar extinction coefficient ($\Delta\varepsilon$, M^{-1} cm^{-1}). The contents of secondary structure were calculated from the CD spectra by the estimation software of Spectra Manager (JASCO, Japan) [5].

3.9. Fluorescence Spectral Measurement

Intrinsic fluorescence spectral measurements were recorded using a fluorescence spectrophotometer F-4600 (Hitachi, Tokyo, Japan). Tris–HCl (50 mmol/L) was used to prepare the protein solution for fluorescence assay. The samples were measured at an emission wavelength of 350 nm to obtain the maximum excitation wavelength and then scanned at this excitation wavelength to record the emission spectra. The E_m and E_x slits were set as 5 nm; scan speed was set as 200 nm/min with a response time of 0.1 s [34,35].

4. Conclusions

Low-intensity ultrasonic processing stimulated the activation of PPO but exposure to high-intensity ultrasonic processing exhibited an inactivation effect on the PPO enzyme. The increasing power under ultrasonic processing induced the polydispersity of the protein, as well as the change of the secondary and tertiary structures of the protein. The loss of α-helical contents after ultrasonic processing led to the reorganization of the secondary structure of the protein. High-intensity ultrasonic processing caused the exposure of Trp residues to a more polar environment, thereby leading to the quenching of fluorescence intensity and subsequently decreasing the suitability of the catalytic center of the enzyme for reaction with substrates.

Supplementary Materials: The following are available online. Figure S1: Elution profile of protein extraction, Figure S2: Lineweaver–Burk equation of purified enzyme using catechol and pyrogallol as substrates. Table S1: Purification of Polyphenol Oxidase (PPO) from orange (*Citrus sinensis* Osbeck).

Author Contributions: Data curation, A.M.; formal analysis, Y.L.; project administration, X.X. and S.P.; software, S.L.; supervision, W.H.; visualization, J.L. and A.I.; writing—original draft, L.Z. (Lijuan Zhu); writing—review and editing, L.Z. (Linhu Zhu).

Funding: This study was gratefully supported by the Fundamental Research Funds for the Central Universities (No. 2662018JC018), National Natural Science Foundation of China (No. 31401507), the project "Research and Demonstration on Key Technologies of New Citrus Juice Processing" (Project No. 2017 yfd0400701) and Hubei Province Technical Innovation Special Major Projects (No. 2018ABA072).

Conflicts of Interest: No conflict of interest exits in the submission of this manuscript, and manuscript is approved by all authors for publication. The submitted work was not carried out in the presence of any personal, professional or financial relationships that could potentially be construed as a conflict of interest.

References

1. Kacem, B.; Matthews, R.F.; Crandall, P.G.; Cornell, J.A. Nonenzymatic browning in aseptically packaged orange juice and orange drinks. Effect of amino acids, deaeration, and anaerobic storage. *J. Food Sci.* **1987**, *52*, 1665–1667. [CrossRef]
2. Landi, M.; Degl'innocenti, E.; Guglielminetti, L.; Guidi, L. Role of ascorbic acid in the inhibition of polyphenol oxidase and the prevention of browning in different browning-sensitive Lactuca sativa var. capitata (L.) and Eruca sativa (Mill.) stored as fresh-cut produce. *J. Sci. Food Agric.* **2013**, *93*. [CrossRef]
3. Huang, N.; Cheng, X.; Hu, W.; Pan, S. Inactivation, aggregation, secondary and tertiary structural changes of germin-like protein in Satsuma mandarine with high polyphenol oxidase activity induced by ultrasonic processing. *Biophys. Chem.* **2015**, *197*, 18–24. [CrossRef]
4. Zhao, G.Y.; Li, B. Studies on the occurrence of non-enzymatic browning during the storage of cloudy apple juice. *Proc. 2007 Int. Conf. Agric. Eng.* **2007**, *32*, 634–643.
5. Cheng, X.; Huang, X.; Liu, S.; Tang, M.; Hu, W.; Pan, S. Characterization of germin-like protein with polyphenol oxidase activity from Satsuma mandarine. *Biochem. Biophys. Res. Commun.* **2014**, *449*, 313–318. [CrossRef]
6. Vamos-Vigyázó, L. Polyphenol Oxidase and Peroxidase in Fruits and Vegetables. *CRC Crit. Rev. Food Sci. Nutr.* **1981**, *15*, 49–127. [CrossRef]
7. Jang, J.H.; Moon, K.D. Inhibition of polyphenol oxidase and peroxidase activities on fresh-cut apple by simultaneous treatment of ultrasound and ascorbic acid. *Food Chem.* **2011**, *124*, 444–449. [CrossRef]
8. Murtaza, A.; Muhammad, Z.; Iqbal, A.; Ramzan, R.; Liu, Y.; Pan, S.; Hu, W. Aggregation and Conformational Changes in Native and Thermally Treated Polyphenol Oxidase From Apple Juice (Malus domestica). *Front. Chem.* **2018**, *6*, 1–10. [CrossRef] [PubMed]
9. Fonteles, T.V.; Costa, M.G.M.; de Jesus, A.L.T.; de Miranda, M.R.A.; Fernandes, F.A.N.; Rodrigues, S. Power ultrasound processing of cantaloupe melon juice: Effects on quality parameters. *Food Res. Int.* **2012**, *48*, 41–48. [CrossRef]
10. Liu, S.; Liu, Y.; Huang, X.; Yang, W.; Hu, W.; Pan, S. Effect of ultrasonic processing on the changes in activity, aggregation and the secondary and tertiary structure of polyphenol oxidase in oriental sweet melon (Cucumis melo var. makuwa Makino). *J. Sci. Food Agric.* **2017**, *97*, 1326–1334. [CrossRef] [PubMed]
11. Illera, A.E.; Sanz, M.T.; Benito-Román, O.; Varona, S.; Beltrán, S.; Melgosa, R.; Solaesa, A.G. Effect of thermosonication batch treatment on enzyme inactivation kinetics and other quality parameters of cloudy apple juice. *Innov. Food Sci. Emerg. Technol.* **2018**, *47*, 71–80. [CrossRef]
12. Yang, C.; Fujita, S.; Nakamura, N. Purification and Characterization of Polyphenol Oxidase from Banana (Musa sapientum L.) Pulp. *J. Agric. Food Chem.* **2000**, *48*, 2732–2735. [CrossRef] [PubMed]
13. Wititsuwannakul, D.; Chareonthiphakorn, N.; Pace, M.; Wititsuwannakul, R. Polyphenol oxidases from latex of Hevea brasiliensis: Purification and characterization. *Phytochemistry* **2002**, *61*, 115–121. [CrossRef]
14. Xu, J.; Zheng, T.; Meguro, S.; Kawachi, S. Purification and characterization of polyphenol oxidase from Henry chestnuts (Castanea henryi). *J. Wood Sci.* **2004**, *50*, 260–265. [CrossRef]
15. Cheng, X.F.; Zhang, M.; Adhikari, B. The inactivation kinetics of polyphenol oxidase in mushroom (Agaricus bisporus) during thermal and thermosonic treatments. *Ultrason. Sonochem.* **2013**, *20*, 674–679. [CrossRef]
16. Ma, H.; Huang, L.; Jia, J.; He, R.; Luo, L.; Zhu, W. Effect of energy-gathered ultrasound on Alcalase. *Ultrason. Sonochem.* **2011**, *18*, 419–424. [CrossRef]

17. Li, R.; Wang, Y.; Hu, W.; Liao, X. Changes in the activity, dissociation, aggregation, and the secondary and tertiary structures of a thaumatin-like protein with a high polyphenol oxidase activity induced by high pressure CO2. *Innov. Food Sci. Emerg. Technol.* **2014**, *23*, 68–78. [CrossRef]
18. Jambrak, A.R.; Mason, T.J.; Lelas, V.; Paniwnyk, L.; Herceg, Z. Effect of ultrasound treatment on particle size and molecular weight of whey proteins. *J. Food Eng.* **2014**, *121*, 15–23. [CrossRef]
19. Bi, X.; Hemar, Y.; Balaban, M.O.; Liao, X. The effect of ultrasound on particle size, color, viscosity and polyphenol oxidase activity of diluted avocado puree. *Ultrason. Sonochem.* **2015**, *27*, 567–575. [CrossRef]
20. Gülseren, I.; Güzey, D.; Bruce, B.D.; Weiss, J. Structural and functional changes in ultrasonicated bovine serum albumin solutions. *Ultrason. Sonochem.* **2007**, *14*, 173–183. [CrossRef] [PubMed]
21. Chandrapala, J.; Zisu, B.; Palmer, M.; Kentish, S.; Ashokkumar, M. Effects of ultrasound on the thermal and structural characteristics of proteins in reconstituted whey protein concentrate. *Ultrason. Sonochem.* **2011**, *18*, 951–957. [CrossRef] [PubMed]
22. Vivian, J.T.; Callis, P.R. Mechanisms of tryptophan fluorescence shifts in proteins. *Biophys. J.* **2001**, *80*, 2093–2109. [CrossRef]
23. Carvalho, A.S.L.; Ferreira, B.S.; Neves-Petersen, M.T.; Petersen, S.B.; Aires-Barros, M.R.; Melo, E.P. Thermal denaturation of HRPA2: pH-dependent conformational changes. *Enzyme Microb. Technol.* **2007**, *40*, 696–703. [CrossRef]
24. Hu, W.; Zhang, Y.; Wang, Y.; Zhou, L.; Leng, X.; Liao, X.; Hu, X. Aggregation and homogenization, surface charge and structural change, and inactivation of mushroom tyrosinase in an aqueous system by subcritical/supercritical carbon dioxide. *Langmuir* **2011**, *27*, 909–916. [CrossRef] [PubMed]
25. Benito-román, Ó.; Sanz, M.T.; Melgosa, R.; de Paz, E.; Escudero, I. Studies of polyphenol oxidase inactivation by means of high pressure carbon dioxide (HPCD). *J. Supercrit. Fluids* **2019**, *147*, 310–321. [CrossRef]
26. Kelly, S.M.; Jess, T.J.; Price, N.C. How to study proteins by circular dichroism. *Biochim. Biophys. Acta - Proteins Proteomics* **2005**, *1751*, 119–139. [CrossRef] [PubMed]
27. Greenfield, N.J. Applications of circular dichroism in protein and peptide analysis. *TrAC - Trends Anal. Chem.* **1999**, *18*, 236–244. [CrossRef]
28. Barteri, M.; Diociaiuti, M.; Pala, A.; Rotella, S. Low frequency ultrasound induces aggregation of porcine fumarase by free radicals production. *Biophys. Chem.* **2004**, *111*, 35–42. [CrossRef] [PubMed]
29. Yu, Z.L.; Zeng, W.C.; Lu, X.L. Influence of ultrasound to the activity of tyrosinase. *Ultrason. Sonochem.* **2013**, *20*, 805–809. [CrossRef] [PubMed]
30. Davis, B.J. Disc Electrophoresis ? II Method and Application to Human Serum Proteins. *Ann. N. Y. Acad. Sci.* **1964**, *121*, 404–427. [CrossRef] [PubMed]
31. Aydemir, T. Partial purification and characterization of polyphenol oxidase from artichoke (Cynara scolymus L.) heads. *Food Chem.* **2004**, *87*, 59–67. [CrossRef]
32. Bradford, M.M. A Rapid and Sensitive Method for the quantitation of Mocrigram Quantities of Protein Utilizing the Principle of Protein-Dye Binding. *Anal. Biochem.* **1976**, *72*, 248–254. [CrossRef]
33. Liu, S.; Murtaza, A.; Liu, Y.; Hu, W.; Xu, X.; Pan, S. Catalytic and Structural Characterization of a Browning-Related Protein in Oriental Sweet Melon (Cucumis Melo var. Makuwa Makino). *Front. Chem.* **2018**, *6*, 1–11. [CrossRef] [PubMed]
34. Murtaza, A.; Iqbal, A.; Linhu, Z.; Liu, Y.; Xu, X.; Pan, S.; Hu, W. Effect of high-pressure carbon dioxide on the aggregation and conformational changes of polyphenol oxidase from apple (Malus domestica) juice. *Innov. Food Sci. Emerg. Technol.* **2019**. [CrossRef]
35. Iqbal, A.; Murtaza, A.; Muhammad, Z.; Elkhedir, A.; Tao, M.; Xu, X. Inactivation, Aggregation and Conformational Changes of Polyphenol Oxidase from Quince (Cydonia oblonga Miller) Juice Subjected to Thermal and High-Pressure Carbon Dioxide Treatment. *Molecules* **2018**, *23*, 1743. [CrossRef] [PubMed]

Sample Availability: Samples of the compounds are available from the authors.

© 2019 by the authors. Licensee MDPI, Basel, Switzerland. This article is an open access article distributed under the terms and conditions of the Creative Commons Attribution (CC BY) license (http://creativecommons.org/licenses/by/4.0/).

Article

Engineering the Enantioselectivity of Yeast Old Yellow Enzyme OYE2y in Asymmetric Reduction of (E/Z)-Citral to (R)-Citronellal

Xiangxian Ying [1,*], Shihua Yu [1], Meijuan Huang [1], Ran Wei [1], Shumin Meng [1], Feng Cheng [1], Meilan Yu [2], Meirong Ying [3], Man Zhao [1] and Zhao Wang [1]

[1] Key Laboratory of Bioorganic Synthesis of Zhejiang Province, College of Biotechnology and Bioengineering, Zhejiang University of Technology, Hangzhou 310014, China; yushihuafx@163.com (S.Y.); meyroline.huang@gmail.com (M.H.); weiranzjut@163.com (R.W.); mengshuminzjut@163.com (S.M.); fengcheng@zjut.edu.cn (F.C.); mzhao@zjut.edu.cn (M.Z.); hzwangzhao@163.com (Z.W.)
[2] College of Life Sciences, Zhejiang Sci-Tech Univeristy, Hangzhou 310018, China; meilanyu@zstu.edu.cn
[3] Grain and Oil Products Quality Inspection Center of Zhejiang Province, Hangzhou 310012, China; hz85672100@163.com
* Correspondence: yingxx@zjut.edu.cn; Tel.: +86-571-88-320-781

Academic Editor: Stefano Serra
Received: 3 March 2019; Accepted: 14 March 2019; Published: 18 March 2019

Abstract: The members of the Old Yellow Enzyme (OYE) family are capable of catalyzing the asymmetric reduction of (E/Z)-citral to (R)-citronellal—a key intermediate in the synthesis of L-menthol. The applications of OYE-mediated biotransformation are usually hampered by its insufficient enantioselectivity and low activity. Here, the (R)-enantioselectivity of Old Yellow Enzyme from *Saccharomyces cerevisiae* CICC1060 (OYE2y) was enhanced through protein engineering. The single mutations of OYE2y revealed that the sites R330 and P76 could act as the enantioselectivity switch of OYE2y. Site-saturation mutagenesis was conducted to generate all possible replacements for the sites R330 and P76, yielding 17 and five variants with improved (R)-enantioselectivity in the (E/Z)-citral reduction, respectively. Among them, the variants R330H and P76C partly reversed the neral derived enantioselectivity from 32.66% e.e. (S) to 71.92% e.e. (R) and 37.50% e.e. (R), respectively. The docking analysis of OYE2y and its variants revealed that the substitutions R330H and P76C enabled neral to bind with a flipped orientation in the active site and thus reverse the enantioselectivity. Remarkably, the double substitutions of R330H/P76M, P76G/R330H, or P76S/R330H further improved (R)-enantioselectivity to >99% e.e. in the reduction of (E)-citral or (E/Z)-citral. The results demonstrated that it was feasible to alter the enantioselectivity of OYEs through engineering key residue distant from active sites, e.g., R330 in OYE2y.

Keywords: asymmetric reduction; citral; citronellal; enantioselectivity; Old Yellow Enzyme; site-saturation mutagenesis; substrate binding mode

1. Introduction

(R)-citronellal is a valuable intermediate for the synthesis of L-menthol through an acidic ene-cyclization and subsequent hydrogenation [1–3]. The potential of (R)-citronellal was also explored for the synthesis of natural vitamin E—a kind of fat-soluble vitamin with relatively high antioxidant ability [4,5]. The commercial Takasago process of (R)-citronellal began with myrcene to form an allylic amine, which underwent asymmetric isomerization in the presence of a 2,2'-bis(diphenylphosphino)-1,1'-binaphthyl(BINAP)-Rh complex and subsequent hydrolysis with acid to give enantiomerically pure (R)-citronellal [6]. In contrast to the three-step asymmetric synthesis from myrcene, the one-step enantioselective reduction of natural citral (the crude mixture of 60% geranial and 40% neral) was a simplified process for the synthesis of (R)-citronellal [7]. The enantioselective hydrogenation of (E/Z)-citral to afford an identical enantiomer remained challenging since the reduction of the geometric isomers geranial and neral by the same catalyst usually yielded the enantiocomplementary products. In organocatalysis, the enantioselective hydrogenation of (E/Z)-citral to yield (R)-citronellal required the use of a dual catalyst system comprising of Pd/BaSO4 and chiral 2-diarylmethylpyrrolidine [8]. However, the obtained (R)-citronellal with 89% e.e. was insufficient for broad industrial applications.

To develop a greener and cost-effective alternative to organocatalysis, Old Yellow Enzymes (OYEs; EC 1.6.99.1) as biocatalysts have been widely investigated, which are capable of catalyzing the C=C bond reduction of α,β-unsaturated compounds such as (E/Z)-citral [9–13]. Past efforts have been made on the discovery of new, improved biocatalysts for suitable enantioselectivity and activity. Bacterial OYEs commonly produced (S)-citronellal from (E/Z)-citral reduction, while the counterparts from yeasts mainly afforded to (R)-enantiomer [10]. Representative yeast OYEs have been well characterized, including OYE2.6 from *Pichia stipites* [9], OYE1 from *Saccharomyces pastorianus*, and OYE2 and OYE3 from *Saccharomyces cerevisiae* [14,15]. So far, the application of OYE-mediated citral reduction still suffers from insufficient enantioselectivity and activity. Protein engineering has emerged as an attractive and powerful strategy for improving enzyme activity and selectivity [16–19]. The circular permutation of OYE1 from *S. pastorianus* yielded the variants exhibiting over an order of magnitude improved catalytic activity [20]. The activity improvement in the protein engineering of yeast OYEs commonly varied by substrate. The variant P295A of OYE1 from *S. pastorianus* showed three- and seven-fold activity for (R)- and (S)-carvone higher than those of wild type enzyme, respectively; however, it had no significant improvement for geranial and neral [21].

With regard to the alteration of OYE enantioselectivity, one of representative examples was the variant W116F of OYE1 from *S. pastorianus* partly reversed the enantioselectivity in the neral reduction from 19% e.e. (S) to 65% e.e. (R) as compared to the wild type [22]. In contrast to the substrate binding mode of the wild type enzyme, the W116F mutation enabled the substrate to bind with a flipped orientation in the active site, and thus reverse the enantioselectivity, while maintaining the same mechanism of trans-hydrogenation of C=C bond [23,24]. W116 is not the sole determinant of enantioselectivity in OYEs, and the enantioselectivity switches seemed to vary by enzyme: Y78, I113, and F247 in OYE2.6 [25]; C26, I69, and H167 in ene reductases YqjM [26]; and W66 and W100 in OYE from *Gluconobacter oxydans* (Gox0502) [27]. The study of enantioselectivity alteration in OYEs rarely use citral as substrate. The latest example was the NADH-dependent cyclohexenone ene reductase from *Zymomonas mobilis* (NCR), in which W66 was critical for controlling the orientation of (E/Z)-citral binding and, thus, the variant W66A/I231A of NCR reversed the geranial derived enantioselectivity from 99% e.e. (S) to 63% e.e. (R) [28].

The study aims to develop the asymmetric reduction of (E/Z)-citral to (R)-citronellal using engineered OYE coupled with formate dehydrogenase for NADH regeneration (Scheme 1). The Old Yellow Enzyme from *S. cerevisiae* CICC1060 (OYE2y) was cloned, overexpressed, and purified, which reduced geranial and neral to citronellal with 82.87% e.e. (R) and 32.66% e.e. (S), respectively. OYE2y was chosen for enantioselectivity alteration since the wild type enzyme showed higher enantioselectivity than OYE1 from *S. pastorianus* and OYE2 and OYE3 from *S. cerevisiae* in the (E/Z)-citral reduction [29]. The key residues for the enantioselectivity of OYE2y were identified through the combination of sequence alignment and single-point mutations. Relying on subsequent site-saturation mutagenesis, the OYE2y variants with double substitutions exhibited full (R)-enantioselectivity in the reduction of (E)-citral or (E/Z)-citral. In addition, the role of key residues and the substrate binding modes were examined via homology modeling and molecular docking.

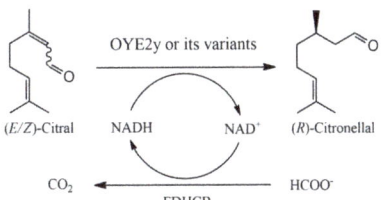

Scheme 1. OYE-mediated asymmetric reduction of (E/Z)-citral to (R)-citronellal coupled with formate dehydrogenase from *Candida boidinii* (FDHCB)-catalyzed NADH regeneration. The reactions were conducted at 37 °C and 200 rpm for 11 h.

2. Results

2.1. OYE2y-Mediated Reduction of (E/Z)-Citral

The yeast Old Yellow Enzyme OYE2y was heterologously expressed in *Echerichia coli* BL21(DE3), and the resulting recombinant OYE2y with N-terminal His tag was purified using affinity chromatography. The enzyme OYE2y with 400 amino acid residues shared the sequence identities of 91.50%, 98.75%, and 81.25% to OYE1 from *S. pastorianus* and OYE2 and OYE3 from *S. cerevisiae* [14], respectively. To investigate the enantioselectivity of OYE2y in citral reduction, the purified OYE2y rather than the whole-cell biocatalyst was used to avoid side reactions. OYE2y accepted NADH or NADPH as coenzyme, and formate/formate dehydrogenase system was used for NADH regeneration in this study. The composition of (E/Z)-citral was determined to contain 58.45% geranial and 41.55% neral. During the first 3 h, the concentration of geranial decreased significantly faster than that of neral. Meanwhile, the concentration of (R)-citronellal increased rapidly, and the *e.e.* value was kept at a higher level of >60% (R) (Figure 1A). Then, the conversion rate of neral turned faster from 3 h to 5 h, resulting in the decreasing of *e.e.* value from 65.02% (R) to 40.26% (R). After 6 h, the conversion rate of neral was nearly parallel to that of geranial, meanwhile the *e.e.* values were kept almost constant. The 11 h reaction was completed with 89.51% yield, giving (R)-citronellal with an *e.e.* value of 38.13% (R). The time course of (E/Z)-citral reduction clearly indicated that the ratio of geranial and neral significantly affected the *e.e.* value of (R)-citronellal, which was consistent with previous observations [7,14]. In addition, the isomerization of geranial and neral occurred under some conditions [9,30]. Thus, the use of freshly prepared geranial and neral with high purity was necessary to study the enantioselectivity of OYEs.

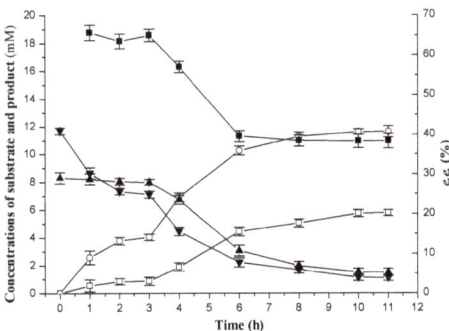

(A)

Figure 1. *Cont.*

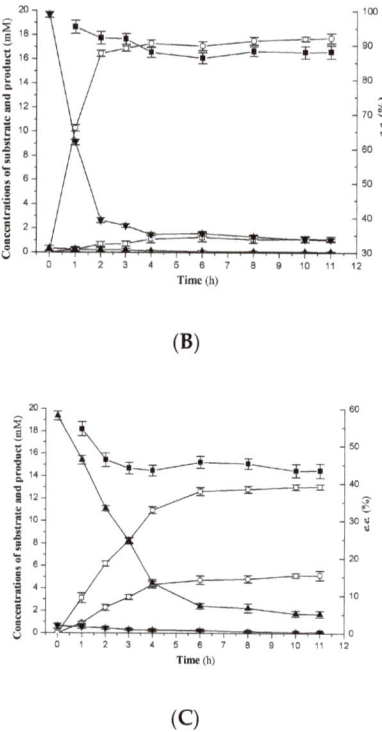

Figure 1. Asymmetric reduction of (E/Z)-citral (**A**), (E)-citral (**B**), and (Z)-citral (**C**) using the purified OYE2y. □, (R)-citronellal; ○, (S)-citronellal; ▲, geranial; ▼, neral; ■, the e.e. value of hydrogenated product. Data present mean values ± SD from three independent experiments.

Considering the high cost of commercial products, geranial and neral with high purity were prepared according to the previous procedure with improvements [31]. Based on the optimized conditions, the yields of (E)-citral and (Z)-citral were increased up to 99.39% and 99.35%, respectively. The obtained (E)-citral sample contained 98.38% geranial and 1.62% neral, and the obtained (Z)-citral sample contained 96.84% neral and 3.16% geranial. When the enzyme OYE2y was tested with newly prepared (E)-citral or (Z)-citral for 4 h, the enantioselectivity of OYE2y stayed at a relatively constant level (Figure 1B,C). The (E)-citral and (Z)-citral-derived e.e. values after 11 h reduction were 82.87% (R) and 32.66% (S), respectively.

2.2. Identification of Key Residues for the Enantioselectivity of OYE2y

It was previously reported that the variant W116F of OYE1 from *S. pastorianus* partly reversed the enantioselectivity in the neral reduction [22]. However, the same substitution at site W117 corresponding to W116 in OYE1 even lowered the e.e. value from 38.13% (R) to 24.01% (R) when (E/Z)-citral was tested as substrate, indicating that the enantioselectivity switch for OYE1 and OYE2y seemed different. Furthermore, the NAD(P)H-dependent enoate reductase (OYE2p) from *S. cerevisiae* YJM1341 was newly discovered for asymmetric reduction of (E/Z)-citral to (R)-citronellal with the e.e. value of 88.8% (R), with four amino acid residues—G13, A59, I289, and H330—in OYE2p different from S13, S59, V289, and R330 in OYE2y (Figure 2). Considering the difference of enantioselectivity between OYE2p and OYE2y, it was expected that S13, S59, V289, and/or R330 might be critical for the enantioselectivity. Then, the single substitutions were conducted to evaluate this expectation. The catalytic performance of the variants S13G, S59A, and V289I was similar to that of OYE2y (Table 1).

The substitution R330 to H significantly increased the (*R*)-enantioselectivity from 38.13% to 86.88% when (*E*/*Z*)-citral was used as substrate. Remarkably, the enantioselectivity was reversed from 32.66% (*S*) to 71.92% (*R*) when (*Z*)-citral was tested. Except for sequence alignment, the identification of key residues was conducted through the single mutation on the randomly-selected residues. Through multiple mutation attempts, the variant P76M was discovered to benefit the (*R*)-enantioselectivity of OYE2y in the citral reduction (Table 1). The substitution P76 to M increased the (*E*/*Z*) citral-derived *e.e.* value from 38.13% (*R*) to 57.60% (*R*), while the enantioselectivity in the reduction of (*Z*)-citral was lowered from 32.66% (*S*) to 4.76% (*S*). Thus, both R330 and P76 were chosen as the targets for subsequent site saturation mutagenesis.

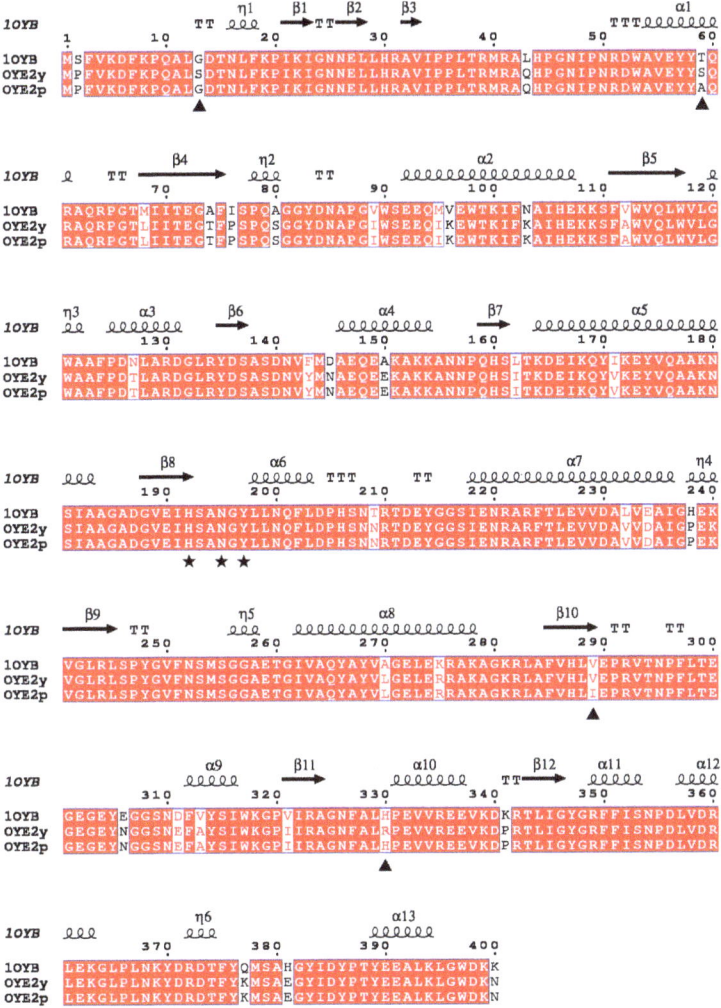

Figure 2. Structure-related sequence alignment between OYE2y and its homologous OYEs. 1OYB: PDB code of OYE1 from *S. pastorianus*; OYE2p: NAD(P)H-dependent enoate reductase from *S. cerevisiae* YJM1341. The secondary structural elements of 1OYB (α-helices, β-strands, T-turns, and η-helices) were indicated above the aligned sequences. The numbering shown was from 1OYB. A red background highlights conserved residues. ▲, the positions where the amino acid residues differed between OYE2y and OYE2p; ★, key residues for catalytic activity. The figure was produced using ESPript 3.0 [32].

Table 1. The catalytic performance of OYE2y and its variants S13G, S59A, P76M, V289I, and R330H [a].

Enzyme	(E)-Citral		(Z)-Citral		(E/Z)-Citral	
	e.e. (%)	Yield (%)	e.e. (%)	Yield (%)	e.e. (%)	Yield (%)
S13G	83.19 ± 1.57 (R)	95.12 ± 2.10	45.29 ± 1.46 (S)	94.60 ± 1.22	37.35 ± 1.63 (R)	96.34 ± 1.87
S59A	80.69 ± 2.21 (R)	90.32 ± 1.54	30.96 ± 1.25 (S)	90.88 ± 2.23	41.66 ± 1.59 (R)	93.48 ± 1.07
V289I	75.18 ± 1.70 (R)	94.88 ± 1.81	37.45 ± 1.55 (S)	91.71 ± 2.10	35.52 ± 2.42 (R)	90.78 ± 0.93
R330H	88.08 ± 1.39 (R)	71.23 ± 0.85	71.92 ± 1.34 (R)	64.12 ± 1.33	86.88 ± 1.36 (R)	52.83 ± 0.75
P76M	86.22 ± 0.75 (R)	80.39 ± 0.64	4.76 ± 1.50 (S)	55.19 ± 0.95	57.60 ± 0.92 (R)	61.53 ± 0.89
OYE2y	82.87 ± 0.98 (R)	92.20 ± 1.07	32.66 ± 1.77 (S)	88.65 ± 1.49	38.13 ± 1.55 (R)	89.51 ± 1.68

[a] Data present mean values ± SD from three independent experiments. (E)-citral contained 98.38% geranial and 1.62% neral, (Z)-citral contained 96.84% neral and 3.16% geranial, and (E/Z)-citral contained 58.45% geranial and 41.55% neral.

2.3. Site-Saturation Mutagenesis of R330 in OYE2y

All R330X variants of OYE2y (X = one of the other 19 amino acids) were successfully expressed in E. coli. After the cells were harvested by centrifugation and then disrupted by ultrasonication, each variant with N-terminal His tag was purified using affinity chromatography. As shown in Figure S1, all 19 variant proteins remained in the soluble fraction, revealing that these substitutions did not decrease the solubility. In comparison with the wild type OYE2y, the R330 variants of OYE2y fell into three categories: R330P without catalytic activity, R330Y with similar (R)-stereoselectivity to OYE2y, and the other 17 variants with improved (R)-stereoselectivity (Table 2). R330P did not retain the yellow color, suggesting that its substitution might deactivate the coenzyme binding. When (E/Z)-citral was tested as substrate, the variants except R330Y and R330P increased the (R)-stereoselectivity but decreased the product yield to some extent. In contrast to the reduction of (E)-citral, those 17 variants showed more significant improvement of (R)-enantioselectivity in the reduction of (Z)-citral. Among them, the variants R330H, R330D, and R330W had superior catalytic performance in terms of activity and enantioselectivity.

Table 2. The catalytic performance of OYE2y and its R330X variants [a].

Enzyme	(E)-Citral		(Z)-Citral		(E/Z)-Citral	
	e.e. (%)	Yield (%)	e.e. (%)	Yield (%)	e.e. (%)	Yield (%)
R330H	88.08 ± 1.39 (R)	71.23 ± 0.85	71.92 ± 1.34 (R)	64.12 ± 1.33	86.88 ± 1.36 (R)	52.83 ± 0.75
R330D	92.42 ± 2.30 (R)	63.12 ± 1.93	73.56 ± 0.51 (R)	36.41 ± 1.94	80.30 ± 1.85 (R)	62.70 ± 0.85
R330W	95.05 ± 1.85 (R)	47.32 ± 1.11	72.19 ± 1.09 (R)	53.38 ± 1.12	79.71 ± 1.03 (R)	70.98 ± 1.15
R330I	87.38 ± 1.11 (R)	31.45 ± 0.40	20.93 ± 1.21 (R)	13.19 ± 0.23	74.52 ± 0.69 (R)	28.17 ± 0.68
R330L	89.64 ± 0.79 (R)	75.14 ± 1.25	45.59 ± 0.75 (R)	55.72 ± 0.59	72.58 ± 1.32 (R)	67.45 ± 1.21
R330F	89.40 ± 1.05 (R)	89.86 ± 1.74	57.68 ± 0.86 (R)	78.39 ± 1.34	72.50 ± 1.46 (R)	80.49 ± 1.69
R330E	89.22 ± 0.61 (R)	80.92 ± 0.80	47.80 ± 1.14 (R)	47.82 ± 1.11	69.42 ± 2.34 (R)	72.46 ± 1.02
R330A	88.10 ± 1.37 (R)	83.10 ± 2.11	51.07 ± 1.23 (R)	49.12 ± 1.63	69.42 ± 1.11 (R)	69.61 ± 0.72
R330T	85.74 ± 2.33 (R)	83.33 ± 2.06	27.97 ± 0.48 (R)	43.83 ± 0.87	69.12 ± 2.32 (R)	46.61 ± 0.29
R330N	92.15 ± 2.49 (R)	90.55 ± 1.85	41.87 ± 0.73 (R)	72.44 ± 1.61	68.19 ± 1.56 (R)	78.89 ± 1.98
R330V	89.66 ± 1.50 (R)	91.10 ± 1.01	24.39 ± 1.55 (R)	65.15 ± 1.69	67.68 ± 1.42 (R)	75.77 ± 1.35
R330S	89.60 ± 1.48 (R)	85.62 ± 1.38	33.95 ± 0.67 (R)	69.67 ± 1.92	64.39 ± 2.05 (R)	71.80 ± 1.57
R330C	87.52 ± 1.56 (R)	80.50 ± 1.47	33.80 ± 1.41 (R)	63.79 ± 1.27	64.35 ± 2.33 (R)	70.16 ± 0.86
R330K	86.18 ± 0.84 (R)	84.82 ± 2.19	5.45 ± 0.59 (R)	63.69 ± 1.51	61.53 ± 1.76 (R)	80.64 ± 0.43
R330Q	84.81 ± 1.41 (R)	90.74 ± 2.55	10.90 ± 0.84 (R)	76.88 ± 0.66	61.17 ± 0.71 (R)	85.23 ± 1.44
R330G	92.39 ± 1.66 (R)	10.96 ± 0.29	5.07 ± 1.22 (R)	8.49 ± 0.15	60.97 ± 1.56 (R)	14.82 ± 0.33
R330M	87.88 ± 1.23 (R)	92.40 ± 1.77	0.37 ± 0.65 (R)	85.18 ± 2.08	58.40 ± 1.08 (R)	86.71 ± 1.37
R330Y	87.03 ± 1.12 (R)	96.61 ± 2.16	40.92 ± 1.53 (S)	88.24 ± 1.89	38.65 ± 1.11 (R)	91.56 ± 1.72
R330P [b]	/	/	/	/	/	/
OYE2y	82.87 ± 0.98 (R)	92.20 ± 1.07	32.66 ± 1.77 (S)	88.65 ± 1.49	38.13 ± 1.55 (R)	89.51 ± 1.68

[a] X represents any of 20 amino acids. Data present mean values ± SD from three independent experiments. (E)-citral contained 98.38% geranial and 1.62% neral, (Z)-citral contained 96.84% neral and 3.16% geranial, and (E/Z)-citral contained 58.45% geranial and 41.55% neral. [b] "/" represents no catalytic activity.

2.4. Site-Saturation Mutagenesis of P76 in OYE2y

Similar to R330X variants, all P76X variants of OYE2y were successfully expressed in *E. coli* and purified (Figure S2). However, the number of the P76 variants in the category without catalytic activity (P76Y, P76Q, P76D, P76E, P76R, P76H, P76F, P76W, and P76K) was obviously greater than that of R330 variants, suggesting that P76 could be also critical for the activity. The category with significantly improved enantioselectivity included P76C, P76S, P76M, P76G, and P76N, whereas the other five variants (P76A, P76V, P76T, P76L, and P76I) showed similar catalytic performance to that of OYE2y (Table 3). Similar to the trend in the R330X variants, higher (*R*)-stereoselectivity of OYE2y variants was accompanied by lower product yields. When (*E/Z*)-citral was used as substrate, the substitution of P76 to C increased the *e.e.* value from 44.13% (*R*) to 69.92% (*R*), but the yield was lowered from 89.51% to 49.65%. Particularly, the *e.e.* value in the reduction of (*Z*)-citral was partly reversed from 32.66% (*S*) to 37.50% (*R*).

Table 3. The catalytic performance of OYE2y and its P76X variants [a].

Enzyme	(*E*)-Citral		(*Z*)-Citral		(*E/Z*)-Citral	
	e.e. (%)	Yield (%)	*e.e.* (%)	Yield (%)	*e.e.* (%)	Yield (%)
P76C	85.39 ± 2.70 (*R*)	65.32 ± 0.76	37.50 ± 1.55 (*R*)	38.54 ± 1.61	69.92 ± 2.46 (*R*)	49.65 ± 1.40
P76S	81.23 ± 1.79 (*R*)	81.84 ± 0.62	10.19 ± 0.73 (*S*)	48.15 ± 0.88	62.80 ± 0.97 (*R*)	64.48 ± 0.90
P76M	86.22 ± 0.75 (*R*)	80.39 ± 0.64	4.76 ± 1.50 (*S*)	55.19 ± 0.95	57.60 ± 1.92 (*R*)	61.53 ± 0.89
P76G	85.52 ± 2.55 (*R*)	83.94 ± 0.78	3.63 ± 1.47 (*S*)	57.54 ± 1.84	53.19 ± 0.81 (*R*)	78.04 ± 1.52
P76N	86.45 ± 1.33 (*R*)	85.36 ± 0.49	13.05 ± 0.67 (*S*)	69.39 ± 1.43	49.59 ± 1.55 (*R*)	72.59 ± 0.77
P76A	86.55 ± 0.91 (*R*)	90.57 ± 1.98	21.21 ± 1.95 (*S*)	88.96 ± 0.87	41.88 ± 0.95 (*R*)	90.43 ± 2.22
P76V	87.54 ± 1.55 (*R*)	92.50 ± 2.51	30.15 ± 1.67 (*S*)	82.30 ± 2.07	40.49 ± 1.01 (*R*)	87.37 ± 1.28
P76T	86.12 ± 1.35 (*R*)	93.79 ± 2.26	28.10 ± 1.69 (*S*)	91.24 ± 2.12	37.05 ± 1.13 (*R*)	91.50 ± 0.94
P76L	84.28 ± 1.62 (*R*)	90.95 ± 1.83	38.01 ± 2.27 (*S*)	88.36 ± 1.46	35.05 ± 1.64 (*R*)	88.46 ± 1.32
P76I	86.10 ± 1.03 (*R*)	94.76 ± 1.74	40.23 ± 1.81 (*S*)	89.95 ± 1.33	34.51 ± 1.17 (*R*)	91.88 ± 2.45
P76Y [b]	/	/	/	/	/	/
P76Q [b]	/	/	/	/	/	/
P76D [b]	/	/	/	/	/	/
P76E [b]	/	/	/	/	/	/
P76R [b]	/	/	/	/	/	/
P76H [b]	/	/	/	/	/	/
P76F [b]	/	/	/	/	/	/
P76W [b]	/	/	/	/	/	/
P76K [b]	/	/	/	/	/	/
OYE2y	82.87 ± 0.98 (*R*)	92.20 ± 1.07	32.66 ± 1.77 (*S*)	88.65 ± 1.49	38.13 ± 1.55 (*R*)	89.51 ± 1.68

[a] X represents one of the other 19 amino acids. Data present mean values ± SD from three independent experiments. (*E*)-citral contained 98.38% geranial and 1.62% neral, (*Z*)-citral contained 96.84% neral and 3.16% geranial, and (*E/Z*)-citral contained 58.45% geranial and 41.55% neral. [b] "/" represents no catalytic activity.

2.5. Evaluation of Double Substitution at Sites P76 and R330 of OYE2y

To investigate the effect of the double substitutions on P76 and R330, we firstly conducted site-directed mutagenesis of P76 to C and R330 to H, D, W, and C, resulting in the four variants (Table 4). The *e.e.* values and yields of the resulting variants were higher than those of the variant P76C, but similar to those of the corresponding R330H, R330D, R330W, and R330C. On the other hand, the other set of double substitutions was created by site-directed mutagenesis of R330 to H and P76 to M, G, and S (Table 4). Among them, the (*E*)-citral-derived *e.e.* values were significantly increased up to >99% (*R*) with relatively higher yields (64.09%~73.88%). In the (*E/Z*)-citral reduction, the variants P76M/R330H, P76M/R330H, and P76M/R330H also exhibited full (*R*)-enantioselectivity despite lower product yields (9.12–15.83%).

Table 4. The catalytic performance of double substitution variants of OYE2y at sites P76 and R330 [a].

Enzyme	(E)-Citral		(Z)-Citral		(E/Z)-Citral	
	e.e. (%)	Yield (%)	e.e. (%)	Yield (%)	e.e. (%)	Yield (%)
P76C	85.39 ± 2.70 (R)	65.32 ± 0.76	37.50 ± 1.55 (R)	38.54 ± 1.61	69.92 ± 2.46 (R)	49.65 ± 1.40
P76C/R330H	88.00 ± 2.04 (R)	78.09 ± 1.16	76.16 ± 0.88 (R)	54.53 ± 0.45	82.32 ± 1.22 (R)	57.74 ± 0.84
P76C/R330D	88.77 ± 1.81 (R)	80.00 ± 1.52	76.45 ± 1.03 (R)	52.23 ± 1.25	81.70 ± 2.18 (R)	63.28 ± 1.36
P76C/R330W	89.53 ± 2.23 (R)	70.30 ± 1.87	70.32 ± 1.16 (R)	56.49 ± 0.68	81.44 ± 1.31 (R)	65.31 ± 1.72
P76C/R330C	87.43 ± 1.37 (R)	90.10 ± 1.96	49.17 ± 2.44 (R)	51.14 ± 0.79	77.37 ± 1.04 (R)	66.19 ± 1.42
R330H	88.08 ± 1.39 (R)	71.23 ± 0.85	71.92 ± 1.34 (R)	64.12 ± 1.33	86.88 ± 1.36 (R)	52.83 ± 0.75
P76M/R330H	>99 (R)	73.88 ± 2.66	75.01 ± 1.59 (R)	13.99 ± 0.23	>99 (R)	15.83 ± 0.45
P76G/R330H	>99 (R)	64.09 ± 1.32	77.60 ± 2.06 (R)	12.64 ± 0.31	>99 (R)	11.61 ± 0.38
P76S/R330H	>99 (R)	64.74 ± 1.57	74.74 ± 1.84 (R)	8.90 ± 0.16	>99 (R)	9.12 ± 0.22
OYE2y	82.87 ± 0.98 (R)	92.20 ± 1.07	32.66 ± 1.77 (S)	88.65 ± 1.49	38.13 ± 1.55 (R)	89.51 ± 1.68

[a] Data present mean values ± SD from three independent experiments. (E)-citral contained 98.38% geranial and 1.62% neral, (Z)-citral contained 96.84% neral and 3.16% geranial, and (E/Z)-citral contained 58.45% geranial and 41.55% neral.

3. Discussion

For the synthesis of (R)-citronellal from OYE-mediated citral reduction, low-cost (E/Z)-citral is the industrial desire in contrast to geranial and neral [9]. However, (E/Z)-citral reduction remains challenging due to limited chemoselectivity and enantioselectivity. On the one hand, the presence of multiple C=C and C=O double bonds of citral makes the whole-cell biocatalyst impossible to avoid side reactions [33]. Thus, the purified enzyme is commonly required as biocatalyst. On the other hand, the hydrogenated products from the geometric isomers geranial and neral are usually enantiocomplementary when the wild type OYE was used as biocatalyst, reducing the reaction's enantioselectivity. The features of (E/Z)-citral reduction make it difficult to implement a high-throughput screening (HTS) method for determining the enantioselectivity of the large variant libraries [34–36]. Typically, the samples must be examined individually by chiral GC, which requires at least thirty minutes per sample. To keep the variant library as minimum as possible, the best strategy for enantioselectivity alteration turns to be site-saturation mutagenesis of individual key residue(s) rather than the HTS-based directed evolution [18,37,38]. Based on the strategy of site-saturated mutagenesis, several groups have successfully increased (R)-enantioselectivity in the (E/Z)-citral reduction, and our study further demonstrated that it was feasible to achieve the full (R)-enantioselectivity in the OYE-mediated (E/Z)-citral reduction through protein engineering.

The R330X and P76X variant libraries yielded 17 and five variants with improved (R)-enantioselectivity, respectively, indicating that the subtle change of structure would significantly affect the enantioselectivity. In the variants R330H and P76C, the amino acid pair R and H possessed electrically charged side group, while the side groups of the amino acid pair P and C was polar and uncharged. It was suggested that the substitution could give priority to amino acid residue(s) with similar side group in terms size, polarity and charge, if no clear structure–function relationship was available. Furthermore, the (Z)-citral-derived e.e. value of R330H (71.92%, R) was much higher than that of OYE2p (26.5%, R) [7], suggesting that the role of S13, S59, and V289 could not be ignored for the (R)-enantioselectivity. For the single substitutions, the (R)-enantioselectivity improvement in the (E/Z)-citral reduction was mainly attributed to the enantioselectivity inversion in the (Z)-citral reduction, meanwhile, the same variant reduced (E)-citral to (R)-citronellal with similar e.e. values. Different from the single substitutions, the double substitutions led to further improved (R)-stereoselectivity in the (E)-citral reduction rather than (Z)-reduction, which was rarely observed in the enantioselectivity alternation of OYEs.

To get detailed insights into the molecular mechanism, the models of OYE2y were created from the OYE1 structure (PDB code: 1OYB) using SWISS-MODEL and molecular docking was conducted using the program AutoDock Vina. The substrate and FMN were docked in silico into the models of OYE2y and its variants. In the model of wild type OYE2y, a conserved H192/N195 pair formed hydrogen bonds with the carbonyl oxygen of α,β-unsaturated carbonyl compounds; a hydride was enantioselectively transferred to the substrate C_β atom from $FMNH_2$; and the Y197 residue provided a proton to the substrate C_α atom as an electron acceptor [17]. From the model of wild type OYE2y, the distances from C_β of neral to the side groups of P76 and R330 were calculated to be 14.92 and 20.84 Å, respectively (Figure 3). Similarly, the distances from C_β of geranial to the side groups of P76 and R330 were calculated to be 14.49 and 21.16 Å, respectively. It was assumed that the residues affecting the enantioselectivity of OYEs might directly interact with the substrate [25]. Our results indicated that the residue distant from active sites, e.g., R330 in OYE2y, could also be pivotal for determining the enantioselectivity of OYEs. Furthermore, substrate modeling into the wild type enzyme revealed two different binding modes for the two isomers geranial and neral, leading to products with different enantioselectivity. The docking analyses of variants P76C, R330H, and P76M/R330 suggested that the reversed enantioselectivity in the neral reduction was due to the flipped binding orientation that placed the opposite face of the alkene above the *si* face of the FMN cofactor [24]. Meanwhile, the same variant reduced (*E*)-citral with preserved (*R*)-enantioselectivity derived from the same binding orientation as wild type OYE2y (Figure 4). Although it was acceptable that enantioselectivity was controlled by tuning the orientation of substrate in the binding sites, how subtle changes control the orientation of substrate binding remains to be unraveled, and X-ray crystallography of the variants in the future study would benefit to clarify the subtle structural differences at the substrate-binding site.

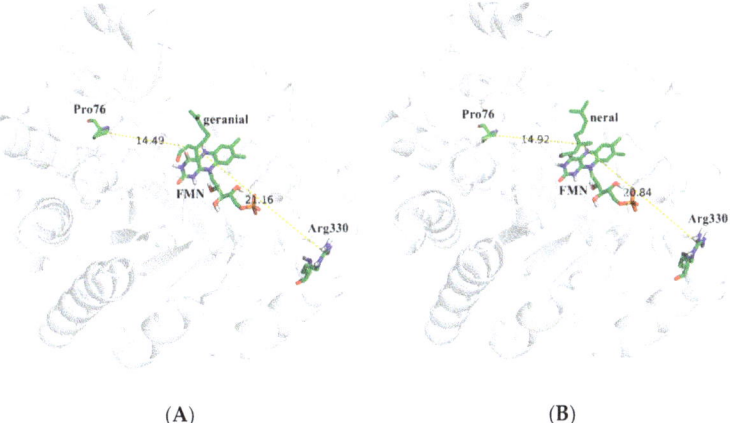

(A) (B)

Figure 3. The residues targeted for site-saturation mutagenesis in the homology model of OYE2y. The model structure of OYE2y was constructed with the crystal structure of OYE1 (PDB code: 1OYB) as template. Distances between the C_β atom of geranial (**A**) and neral (**B**) and side chains of P76 and R330 were determined. Green, carbon atom; blue, nitrogen atom; tangerine, oxygen atom; white, hydrogen atom; orange, phosphorus atom.

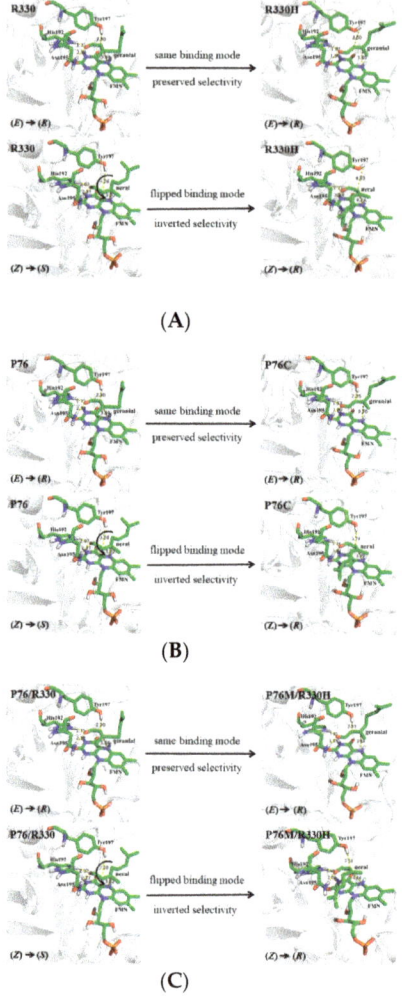

Figure 4. The binding modes of citral isomers in OYE2y and its variants R330H (**A**), P76C (**B**), and P76M/R330H (**C**) leading to either (*R*)- or (*S*)-citronellal. The catalytic residues H192, N195, Y197, and the prosthetic group FMN were depicted. Green, carbon atom; blue, nitrogen atom; tangerine, oxygen atom; white, hydrogen atom; orange, phosphorus atom.

4. Materials and Methods

4.1. Organisms and Chemicals

The organism *S. cerevisiae* CICC1060 was purchased from the China Center of Industrial Culture Collection (CICC, Beijing, China). *S. cerevisiae* CICC1060 was cultured with YPD medium (tryptone 20 g/L, yeast extract 10 g/L, and glucose 20 g/L) at 30 °C for 24 h. The pEASY-E1 expression vector from TransGen Biotech Co., Ltd (Beijing, China) was used for overexpression of the enzyme OYE2y, and the *E. coli* strain BL21(DE3) was used as the host. *E. coli* cultures were grown routinely in Luria Bertani (LB) medium at 37 °C for 12 h.

The standards (*S*)-citronellal, (*R*)-citronellal, and (*S/R*)-citronellal were obtained from Sigma-Aldrich (Shanghai) Trading Co., Ltd. (Shanghai, China). Other chemicals of analytical grade were purchased from Sangon Biotech Co. Ltd (Shanghai, China) or Shanghai Jingchun Reagent Co., Ltd (Shanghai, China). The site-directed mutagenesis kit and the restriction enzyme *Dpn* I were obtained from Vazyme Biotech Co., Ltd. (Nanjing, China). KOD DNA polymerase was purchased from TransGen Biotech Co., Ltd (Beijing, China). The Ni-NTA-HP resin column for protein purification was obtained from GE Healthcare Life Sciences (Shanghai, China).

4.2. Preparation of (Z)-Citral and (E)-Citral

(Z)-citral and (E)-citral were prepared by a modification of the procedure described previously [38]. Activated MnO_2 (1.15 g) was added into a 50-mL three-necked bottom round flask which atmosphere was replaced with N_2. One-hundred milligrams of geraniol or nerol was dissolved in 16 mL dry hexane and was charged in the flask to initiate the alcohol oxidation. The reaction was maintained at 0 °C and 450 rpm for 6 h. Then, the reaction solution was filtered through a filter paper and hexane in the filtrate was removed by vacuum evaporation at 45 °C. Finally, an aliquot of the collected product (E)-citral or (Z)-citral was dissolved in ethyl acetate and subjected to the analyses of gas chromatography (GC) and gas chromatography–mass spectrometry (GC–MS).

4.3. Cloning, Expression, and Purification of OYE2y

The gene encoding OYE2y was PCR-amplified from the genomic DNA of *S. cerevisiae* CICC1060 using a set of primers: Forward, 5′-ATGCCATTTGTTAAGGACTTTAAGCCAC-3′; Reverse, 5′-TTAATTTTTGTCCCAACCGAGTTTTAGAGC-3′. The conditions for PCR amplification of the *oye2y* gene were 94 °C for 2 min for initial denaturalization, 30 cycles of 94 °C for 30 s, 57 °C for 30 s, 72 °C for 80 s, and 72 °C for 10 min for the final extension.

Following the procedure of expression and purification of ReBDH [39], the PCR products were purified and then ligated with the expression vector pEASY-E1 through the AT ligation strategy. The recombinant plasmid harboring the *oye2y* gene, designated as pEASY-E1-*oye2y*, was verified by DNA sequencing (Sangon Biotech, Shanghai, China) and then transformed into *E. coli* BL21 (DE3) competent cells, resulting in the recombinant strain *E. coli* BL21(DE3)/pEASY-E1-*oye2y*. The recombinant cells containing pEASY-E1-*oye2y* were grown in the LB medium with 100 μg/mL ampicillin at 37 °C and 200 rpm until the OD_{600} was 0.6–0.8, and then 0.2 mM isopropyl β-D-1-thiogalactopyranoside (IPTG) was supplemented to initiate the induction at 23 °C and 160 rpm. After 12 h of growth, recombinant *E. coli* cells were harvested by centrifugation and further washed using 50 mM Tris-HCl buffer (pH 8.0). The cells were disrupted through ultrasonication for 10 min, and the cell debris and cell lysate were removed by centrifugation to result in a clear cell extract. The crude cell extracts containing OYE2y was applied to a Ni-NTA chelating affinity column equilibrated with the binding buffer (5 mM imidazole and 300 mM NaCl dissolved in 50 mM Tris-HCl, pH 8.0). Unbound proteins were washed off by the application of the binding buffer. The recombinant OYE2y was eluted with 100 mM imidazole in 50 mM Tris-HCl (pH 8.0), desalted with 50 mM Tris-HCl buffer (pH 8.0) by ultrafiltration and then stored at −20 °C for further study.

4.4. Construction of OYE2y Variants by Site-Directed Mutagenesis

The variants with single or double substitutions were constructed by site-directed mutagenesis according to the QuikChange Mutagenesis Kit. PCR amplification to introduce substitution was performed in 50 μL of standard PCR mixture with 50 ng of template plasmid DNA and 15 pmol each of the appropriate set of primers using the following temperature cycle; 30 s at 95 °C, followed by 30 cycles of 95 °C for 15 s, appropriate annealing temperature (55~61 °C) for 15 s, and 72 °C for 80 s, and the final extension of 5 min at 72 °C. The plasmid pEASY-E1-*oye2y* was used as template DNA in the single substitution, while the creation of double mutants was based on the generated OYE2y P76C or R330H gene as the template. The primer sets for single or double substitutions were shown in Tables S1 and S2. The amplified PCR fragments were digested with the restriction enzyme *Dpn* I at 37 °C for 1 h, and then the digested DNA was directly introduced into *E. coli* strain BL21(DE3).

Each constructed plasmid was confirmed by sequencing. Expression and purification of the resulting OYE2y variants were conducted using the same procedure as OYE2y.

4.5. Homology Modeling and Molecular Docking

Using the crystal structure of OYE1 from *S. pastorianus* (PDB number: 1OYB) as a template [40], the structural model of OYE2y was obtained by a homology modeling strategy [41,42]. Molecular docking simulation was performed by AutoDock Vina [43] when the binding package was set at a distance of 15 Å from FMN N_5 atom. (*E*)-citral or (*Z*)-citral acted as a ligand to molecular docking with OYE2y, and the calculation of geometric parameters and ligand structure was performed by ChemBioDraw 12.0 (CambridgeSoft, Cambridge, MA, USA). To make the results more accurate, 100 consecutive runs were performed and the highest ranked score from each run was used to calculate the average score of each flexible ligand configuration. The optimal configuration and the resulting substrate–enzyme complexes were further processed using the PYMOL software [44]. The candidate complexes were acceptable when they met both criteria: (1) the substrate carbonyl oxygen should be capable of forming hydrogen bonds with the side chains of both H192 and N195 and (2) the distance between the FMN N_5 atom and the β-unsaturated carbon of the substrate molecule was be in an appropriate range from 3.5 Å to 4.1 Å [22,45,46].

4.6. Asymmetric Reduction of Citral Mediated by OYE2y or Its Variants

The reaction mixture (1 mL) contained 50 mM PIPES buffer (pH 7.0), 20 mM substrate, 1 mg OYE2y or its variant, 0.6 U formate dehydrogease from *Candida boidinii* (FDHCB), and 100 mM sodium formate, 0.96 mM NAD^+. The substrate included (*Z*)-citral, (*E*)-citral, or (*E*/*Z*)-citral, which stock solution was 200 mM substrate in isopropanol. The overexpression and purification of FDHCB were conducted according to the procedure described previously [39]. The reaction was conducted at 30 °C and 200 rpm for 11 h, unless otherwise specified. The reaction mixture was centrifuged to remove the cells, and the resulting supernatant was extracted with equal volume of ethyl acetate at 30 °C and 200 rpm for 2 h. Finally, the solvent phase was collected, dried over anhydrous sodium sulfate and subjected to the analyses of GC and GC–MS.

4.7. Analyses of GC and GC–MS

(*Z*)-citral, (*E*)-citral, (*S*)-citronellal, and (*R*)-citronellal were determined by GC (Agilent 6890N) equipped with an FID detector and chiral capillary BGB-174 column (BGB Analytik, Böckten, Switzerland, 30 m × 250 µm × 0.25 µm). The flow rate and split ratio of N_2 as the carrier gas were set as 1.38 mL/min and 1:100, respectively. Both injector and detector were kept at 250 °C. The column temperature program was listed as follows; initial temperature of 90 °C for 25 min, 20 °C/min ramp to 150 °C for 3 min, and 30 °C/min ramp to 180 °C for 3 min. The injection volume was 1 µL. The retention times of (*S*)-citronellal, (*R*)-citronellal, (*Z*)-citral, and (*E*)-Citral were 22.459 min, 23.067 min, 29.164 min and 30.398 min, respectively (Figure S3).

(*S*)-citronellal, (*R*)-citronellal, (*Z*)-citral, and (*E*)-Citral were validated through GC–MS analysis (Figure S4). The GC–MS analysis (Agilent7890A/5975C, Agilent Technologies Inc., Santa Clara, CA, USA) comprised the following parameters; auxiliary heating zone temperature, 250 °C; MS quadrupole temperature, 150 °C; ion source temperature, 230 °C; scan quality range, 30–500 amu; emission current, 200 µA; and electron energy, 70 eV.

4.8. Nucleotide Sequence Accession Number

The gene encoding OYE2y has been deposited in the GenBank database under the accession numbers of MK372229.

5. Conclusions

In summary, significant increase of (R)-enantioselectivity in the (E/Z)-citral reduction was achieved by saturation mutagenesis of P76 and R330 in OYE2y. Remarkably, the variants P76M/R330H, P76G/R330H, and P76S/R330H exhibited full (R)-enantioselectivity in the reduction of (E)-citral or (E/Z)-citral. The variants with improved (R)-enantioselectivity usually came along with lower catalytic activities, indicating that the sites P76 and R330 were important for enantioselectivity as well as activity. In contrast to P76, R330 was relatively distant from active sites and its substitutions brought more beneficial impacts on enantioselectivity. Our results proved that it was reasonable to alter the enantioselectivity of OYE2y by saturation mutagenesis of key residue distant from active sites.

Supplementary Materials: The following are available online at http://www.mdpi.com/1420-3049/24/6/1057/s1, Table S1: The primer information of site saturation mutation of P76 in OYE2y; Table S2: The primer information of site saturation mutation of R330 in Oye2y; Figure S1: SDS-PAGE (12%) analysis of the purified OYE2y R330X variants; Figure S2: SDS-PAGE (12%) analysis of the purified OYE2y P76X variants; Figure S3: Gas chromatography analysis for standards (S)-citronellal (22.459 min), (R)-citronellal (23.067 min), (Z)-citral (29.164 min), and (E)-citral (30.398 min); Figure S4: Gas chromatography–mass spectrometry analysis for (S)-citronellal (A), (R)-citronellal (B), (Z)-citral (C), and (E)-citral (D) in the asymmetric reduction of (E/Z)-citral.

Author Contributions: Conceptualization, X.Y.; Data Curation, X.Y., M.Y. (Meilan Yu), and M.Y. (Meirong Ying); Formal Analysis, X.Y., M.Y. (Meilan Yu), and M.Y. (Meirong Ying); Funding Acquisition, X.Y.; Investigation, S.Y., M.H., R.W., S.M., F.C., and M.Z.; Supervision, Z.W.; Writing—Original Draft, X.Y.

Funding: This work was supported by the Natural Science Foundation of Zhejiang Province, China (No. LY17B020012).

Conflicts of Interest: The authors declare no conflict of interest.

References

1. Itoh, H.; Maeda, H.; Yamada, S.; Hori, Y.; Mino, T.; Sakamoto, M. Kinetic resolution of citronellal by chiral aluminum catalysts: L-menthol synthesis from citral. *Org. Chem. Front.* **2014**, *1*, 1107–1115. [CrossRef]
2. Lenardão, E.J.; Botteselle, G.V.; de Azambuja, F.; Perin, G.; Jacob, R.G. Citronellal as key compound in organic synthesis. *Tetrahedron* **2007**, *63*, 6671–6712. [CrossRef]
3. Nie, Y.; Chuah, G.-K.; Jaenicke, S. Domino-cyclisation and hydrogenation of citronellal to menthol over bifunctional Ni/Zr-Beta and Zr-beta/Ni-MCM-41 catalysts. *Chem. Commun.* **2006**, 790–792. [CrossRef] [PubMed]
4. Coffen, D.L.; Cohen, N.; Pico, A.M.; Schmid, R.; Sebastian, M.J.; Wang, F. A microbial lipase based stereoselective synthesis of (d)-α-tocopherol from (R)-citronellal and (S)-(6-hydroxy-2,5,7,8-tetramethylchroman-2-yl)acetic acid. *Heterocycles* **1994**, *39*, 527–552. [CrossRef]
5. Eggersdorfer, M.; Laudert, D.; Létinois, U.; McClymont, T.; Medlock, J.; Netscher, T.; Bonrath, W. One hundred years of vitamins—a success story of the natural sciences. *Angew. Chem. Int. Ed.* **2012**, *51*, 12960–12990. [CrossRef] [PubMed]
6. Tani, K.; Yamagata, T.; Akutagawa, S.; Kumobayashi, H.; Taketomi, T.; Takaya, H.; Miyashita, A.; Noyori, R.; Otsuka, S. Highly enantioselective isomerization of prochiral allylamines catalyzed by chiral diphosphine Rhodium(I) complexes: Preparation of optically active enamines. *J. Am. Chem. Soc.* **1984**, *106*, 5208–5217. [CrossRef]
7. Zheng, L.; Lin, J.; Zhang, B.; Kuang, Y.; Wei, D. Identification of a yeast old yellow enzyme for highly enantioselective reduction of citral isomers to (R)-citronellal. *Bioresour. Bioprocess.* **2018**, *5*, 9. [CrossRef]
8. Maeda, H.; Yamada, S.; Itoh, H.; Hori, Y. A dual catalyst system provides the shortest pathway for L-menthol synthesis. *Chem. Commun.* **2012**, *48*, 1772–1774. [CrossRef] [PubMed]
9. Bougioukou, D.J.; Walton, A.Z.; Stewart, J.D. Towards preparative-scale, biocatalytic alkene reductions. *Chem. Comm.* **2010**, *46*, 8558–8560. [CrossRef] [PubMed]
10. Müller, A.; Hauer, B.; Rosche, B. Enzymatic reduction of the α,β-unsaturated carbon bond in citral. *J. Mol. Catal. B Enzym.* **2006**, *38*, 126–130. [CrossRef]
11. Toogood, H.S.; Gardiner, J.M.; Scrutton, N.S. Biocatalytic reductions and chemical versatility of the old yellow enzyme family of flavoprotein oxidoreductases. *ChemCatChem* **2010**, *2*, 892–914. [CrossRef]

12. Toogood, H.S.; Scrutton, N.S. New developments in 'ene'-reductase catalysed biological hydrogenations. *Curr. Opin. Chem. Biol.* **2014**, *19*, 107–115. [CrossRef]
13. Toogood, H.S.; Scrutton, N.S. Discovery, characterization, engineering, and applications of ene-reductases for industrial biocatalysis. *ACS Catal.* **2018**, *8*, 3532–3549. [CrossRef]
14. Müller, A.; Hauer, B.; Rosche, B. Asymmetric alkene reduction by yeast old yellow enzymes and by a novel *Zymomonas mobilis* reductase. *Biotechnol. Bioeng.* **2007**, *98*, 22–29. [CrossRef]
15. Brenna, E.; Gatti, F.G.; Monti, D.; Parmeggiani, F.; Serra, S. Stereochemical outcome of the biocatalysed reduction of activated tetrasubstituted olefins by old yellow enzymes 1–3. *Adv. Synth. Catal.* **2012**, *354*, 105–112. [CrossRef]
16. Amato, E.D.; Stewart, J.D. Applications of protein engineering to members of the old yellow enzyme family. *Biotechnol. Adv.* **2015**, *33*, 624–631. [CrossRef]
17. Kataoka, M.; Miyakawa, T.; Shimizu, S.; Tanokura, M. Enzymes useful for chiral compound synthesis: Structural biology, directed evolution, and protein engineering for industrial use. *Appl. Microbiol. Biotechnol.* **2016**, *100*, 5747–5757. [CrossRef]
18. Li, R.; Wijma, H.J.; Song, L.; Cui, Y.; Otzen, M.; Tian, Y.; Du, J.; Li, T.; Niu, D.; Chen, Y.; et al. Computational redesign of enzymes for regio- and enantioselective hydroamination. *Nature Chem. Biol.* **2018**, *14*, 664–670. [CrossRef]
19. Ying, X.; Zhang, J.; Wang, C.; Huang, M.; Ji, Y.; Cheng, F.; Yu, M.; Wang, Z.; Ying, M. Characterization of a carbonyl reductase from *Rhodococcus erythropolis* WZ010 and its variant Y54F for asymmetric synthesis of (S)-N-Boc-3-hydroxypiperidine. *Molecules* **2018**, *23*, 3117. [CrossRef]
20. Daugherty, A.B.; Govindarajan, S.; Lutz, S. Improved biocatalysts from a synthetic circular permutation library of the flavin-dependent oxidoreductase old yellow enzyme. *J. Am. Chem. Soc.* **2013**, *135*, 14425–14432. [CrossRef]
21. Quertinmont, L.T.; Lutz, S. Cell-free protein engineering of Old Yellow Enzyme 1 from *Saccharomyces pastorianus*. *Tetrahedron* **2016**, *72*, 7282–7287. [CrossRef]
22. Padhi, S.K.; Bougioukou, D.J.; Stewart, J.D. Site-saturation mutagenesis of tryptophan 116 of *Saccharomyces pastorianus* old yellow enzyme uncovers stereocomplementary variants. *J. Am. Chem. Soc.* **2009**, *131*, 3271–3280. [CrossRef]
23. Brenna, E.; Crotti, M.; Gatti, F.G.; Monti, D.; Parmeggiani, F.; Powell, R.W., III; Santangelo, S.; Stewart, J.D. Opposite enantioselectivity in the bioreduction of (Z)-β-aryl-bcyanoacrylates mediated by the tryptophan 116 mutants of old yellow enzyme 1: Synthetic approach to (R)- and (S)-β-aryl-glactams. *Adv. Synth. Catal.* **2015**, *357*, 1849–1860. [CrossRef]
24. Pompeu, Y.A.; Sullivan, B.; Stewart, J.D. X-ray crystallography reveals how subtle changes control the orientation of substrate binding in an alkene reductase. *ACS Catal.* **2013**, *3*, 2376–2390. [CrossRef]
25. Walton, A.Z.; Sullivan, B.; Patterson-Orazem, A.C.; Stewart, J.D. Residues controlling facial selectivity in an alkene reductase and semirational alterations to create stereocomplementary variants. *ACS Catal.* **2015**, *4*, 2307–2318. [CrossRef]
26. Rüthlein, E.; Classen, T.; Dobnikar, L.; Schölzel, M.; Pietruszka, J. Finding the selectivity switch—A rational approach towards stereocomplementary variants of the ene reductase YqjM. *Adv. Synth. Catal.* **2015**, *357*, 1775–1786. [CrossRef]
27. Yin, B.; Deng, J.; Lim, L.; Yuan, Y.A.; Wei, D. Structural insights into stereospecific reduction of α, β-unsaturated carbonyl substrates by old yellow enzyme from *Gluconobacter oxydans*. *Biosci. Biotechnol. Biochem.* **2015**, *79*, 410–421. [CrossRef]
28. Kress, N.; Rapp, J.; Hauer, B. Enantioselective reduction of citral isomers in NCR ene reductase: Analysis of an active site mutant library. *ChemBioChem* **2017**, *18*, 717–720. [CrossRef]
29. Hall, M.; Stueckler, C.; Hauer, B.; Stuermer, R.; Friedrich, T.; Breuer, M.; Kroutil, W.; Faber, K. Asymmetric bioreduction of activated C=C bonds using *Zymomonas mobilis* NCR enoate reductase and old yellow enzymes OYE 1-3 from yeasts. *Eur. J. Org. Chem.* **2008**, 1511–1516. [CrossRef]
30. Wolken, W.A.M.; ten Have, R.; van der Werf, M.J. Amino acid-catalyzed conversion of citral: *Cis-trans* isomerization and its conversion into 6-methyl-5-hepten-2-one and acetaldehyde. *J. Agric. Food Chem.* **2000**, *48*, 5401–5405. [CrossRef]
31. Tsuboi, S.; Ishii, N.; Sakai, T.; Tari, I.; Utaka, M. Oxidation of alcohols with electrolytic manganese dioxide. Its application fro the synthesis of insect pheromones. *Bull. Chem. Soc. Jpn.* **1990**, *63*, 1888–1893. [CrossRef]

32. Robert, X.; Gouet, P. Deciphering key features in protein structures with the new ENDscript server. *Nucleic Acids Res.* **2014**, *42*, W320–W324. [CrossRef]
33. Hall, M.; Hauer, B.; Stuermer, R.; Kroutil, W.; Faber, K. Asymmetric whole-cell bioreduction of an α,β-unsaturated aldehyde (citral): Competing prim-alcohol dehydrogenase and C-C lyase activities. *Tetrahedron: Asymmetry* **2006**, *17*, 3058–3062. [CrossRef]
34. Cheng, F.; Tang, X.; Kardashliev, T. Transcription factor-based biosensors in high-throughput screening: Advances and applications. *Biotechnol. J.* **2018**, *13*, 1700648. [CrossRef]
35. Cheng, F.; Zhu, L.; Schwaneberg, U. Directed evolution 2.0: Improving and deciphering enzyme properties. *Chem. Commun.* **2015**, *51*, 9760–9772. [CrossRef]
36. Deng, J.; Yao, Z.; Chen, K.; Yuan, Y.A.; Lin, J.; Wei, D. Towards the computational design and engineering of enzyme enantioselectivity: A case study by a carbonyl reductase from *Gluconobacter oxydans*. *J. Biotechnol.* **2016**, *217*, 31–40. [CrossRef]
37. Sullivan, B.; Walton, A.Z.; Stewart, J.D. Library construction and evaluation for site saturation mutagenesis. *Enzyme Microb. Technol.* **2013**, *53*, 70–77. [CrossRef]
38. Valetti, F.; Gilardi, G. Improvement of biocatalysts for industrial and environmental purposes by saturation mutagenesis. *Biomolecules* **2013**, *3*, 778–811. [CrossRef]
39. Yu, M.; Huang, M.; Song, Q.; Shao, J.; Ying, X. Characterization of a (2R,3R)-2,3-butanediol dehydrogenase from *Rhodococcus erythropolis* WZ010. *Molecules* **2015**, *20*, 7156–7173. [CrossRef]
40. Fox, K.M.; Karplus, P.A. Old yellow enzyme at 2 Å resolution: Overall structure, ligand binding, and comparison with related flavoproteins. *Structure* **1994**, *2*, 1089–1105. [CrossRef]
41. Arnold, K.; Bordoli, L.; Kopp, J.; Schwede, T. The SWISS-MODEL workspace: A web-based environment for protein structure homology modelling. *Bioinformatics* **2006**, *22*, 195–201. [CrossRef]
42. Schwede, T.; Kopp, J.; Guex, N.; Peitsch, M.C. SWISS-MODEL: An antomated protein homology-modeling server. *Nucleic Acids Res.* **2003**, *31*, 3381–3385. [CrossRef]
43. Trott, O.; Olson, A.J. AutoDock Vina: Improving the speed and accuracy of docking with a new scoring function, efficient optimization, and multithreading. *J. Comput. Chem.* **2010**, *31*, 455–461. [CrossRef]
44. Seeliger, D.; de Groot, B.L. Ligand docking and binding site analysis with PyMOL and Autodock/Vina. *J. Comput. Aided Mol. Des.* **2010**, *24*, 417–422. [CrossRef]
45. Breithaupt, C.; Strassner, J.; Breitinger, U.; Huber, R.; Macheroux, P.; Schaller, A.; Clausen, T. X-Ray structure of 12-oxophytodienoate reductase 1 provides structural insight into substrate binding and specificity within the family of OYE. *Structure* **2001**, *9*, 419–429. [CrossRef]
46. Fraaije, M.W.; Mattevi, A. Flavoenzymes: Diverse catalysts with recurrent features. *Trends Biochem. Sci.* **2000**, *25*, 126–132. [CrossRef]

Sample Availability: Samples of the compounds are not available from the authors

© 2019 by the authors. Licensee MDPI, Basel, Switzerland. This article is an open access article distributed under the terms and conditions of the Creative Commons Attribution (CC BY) license (http://creativecommons.org/licenses/by/4.0/).

Article

Antifungal Activity against *Botrytis cinerea* of 2,6-Dimethoxy-4-(phenylimino)cyclohexa-2,5-dienone Derivatives

Paulo Castro [1,*], Leonora Mendoza [1], Claudio Vásquez [2], Paz Cornejo Pereira [1], Freddy Navarro [1], Karin Lizama [1], Rocío Santander [3] and Milena Cotoras [1]

1. Laboratorio de Micología, Facultad de Química y Biología, Universidad de Santiago de Chile, Avenida Libertador Bernardo O'Higgins 3363, Santiago 518000, Chile; leonora.mendoza@usach.cl (L.M.); paz.cornejo@usach.cl (P.C.P.); freddy.navarro@usach.cl (F.N.); karin.lizama@usach.cl (K.L.); milena.cotoras@usach.cl (M.C.)
2. Laboratorio de Microbiología Molecular, Departamento de Biología, Facultad de Química y Biología, Universidad de Santiago de Chile, Santiago 518000, Chile; claudio.vasquez@usach.cl
3. Departamento de Ciencias del Ambiente, Facultad de Química y Biología, Universidad de Santiago de Chile, Casilla 40 Correo 33, Santiago 518000, Chile; rocio.santanderm@usach.cl
* Correspondence: paulo.castro@usach.cl; Tel.: +56-2-2718-1063

Academic Editor: Stefano Serra
Received: 20 December 2018; Accepted: 12 February 2019; Published: 15 February 2019

Abstract: In this work the enzyme laccase from *Trametes versicolor* was used to synthetize 2,6-dimethoxy-4-(phenylimino)cyclohexa-2,5-dienone derivatives. Ten products with different substitutions in the aromatic ring were synthetized and characterized using ^{1}H- and ^{13}C-NMR and mass spectrometry. The 3,5-dichlorinated compound showed highest antifungal activity against the phytopathogen *Botrytis cinerea*, while the *p*-methoxylated compound had the lowest activity; however, the antifungal activity of the products was higher than the activity of the substrates of the reactions. Finally, the results suggested that these compounds produced damage in the fungal cell wall.

Keywords: *Botrytis cinerea*; antifungal activity; laccase; 2,6-dimethoxy-4-(phenylimino)cyclohexa-2,5-dienone derivatives

1. Introduction

Botrytis cinerea is a phytopathogenic fungus promoted by the presence of free surface water or high relative humidity and causing significant crop losses in a wide variety of plant species [1]. Regarding the control, methods aiming to reduce humidity can be combined to help decrease this disease, in addition to chemical fungicides or biocontrol treatments [1]. Chemical control is the most common way to manage *B. cinerea*, mainly using synthetic compounds [1]. The restriction of this type of control becomes necessary to reduce the impact on the environment [2] and to avoid the acquired resistance to botrycides [3–7]. For this reason, the development of new antifungal compounds is essential. Natural products can be a good alternative to commercial fungicides [8,9]. For instance, phenolic compounds, terpenoids, nitrogen-containing compounds, and aliphatic compounds isolated from plants have shown antifungal activities [10–12]. Additionally, new antifungal compounds against *B. cinerea* derived of natural products have been synthesized, such as derivatives of natural stilbene resveratrol [13], chlorophenyl derivatives [14], or different clovanes [15].

Several phenolic metabolites found in grape pomace have shown low antifungal activity against *B. cinerea* [16], therefore, it is possible to increase the biological activity of phenolic compounds using the enzyme laccase [17]. These enzymes (benzenediol: oxygen oxidoreductase, EC 1.10.3.2) belong to the oxidase group, and they are also used for cleaner industrial application [18]. Laccases are also known as multicopper oxidases, they belong to the family of copper-containing phenol oxidases [19] and can oxidize a diversity of compounds, e.g., phenolic and nonphenolic compounds [18]. Aromatic compounds can produce reactive radical intermediates, which undergo self-coupling reactions, thus forming different dimers and trimers [20–24]. This enzyme has been previously used to improve the activity of antibiotics [25,26]. On the other hand, the synthesis of a heterodimeric compound (2,6-dimethoxy-4-(phenylimino)cyclohexa-2,5-dienone) by the laccase-mediated coupling reaction between syringic acid and aniline was reported, this compound showed an antifungal effect against B. cinerea with an EC_{50} value of 0.14 mM [27].

Antifungal compounds have shown several inhibition mechanisms related to the molecular structure. For instance, the resveratrol derivative (E)-3,5-dimethoxy-β-(2-furyl)-styrene cause cell membrane damage against B. cinerea [13]. Phenylpyrroles induce morphological alterations of germ tubes [28]. Fungicides such as dinocap and fuazinam have been described as uncouplers of oxidative phosphorylation [29,30] and fungicides, like dicloran, cloroneb, and etazol, affect cell wall synthesis [28].

This work aimed to determine the antifungal activity against B. cinerea and the effect on the cell wall integrity of ten 2,6-dimethoxy-4-(phenylimino)cyclohexa-2,5-dienone derivatives (3a–j) obtained by reaction of syringic acid (1) with substituted anilines (2a–j). To analyze the effect of the carboxylic group in these laccase-catalyzed reactions, syringaldehyde was used instead syringic acid and the reaction product was characterized.

2. Results and Discussion

2.1. Laccase-Mediated Synthesis of 2,6-Dimethoxy-4-(phenylimino)cyclohexa-2,5-dienone Derivatives

In this work, laccase catalyzed reactions using 1 and 2a–j were carried out. It has been previously reported that using laccases from different fungal sources (Trametes sp. and Rhizoctonia praticola), catalyze reactions between phenolic compounds and anilines, heterodimeric compounds are formed [25,26,31,32], similarly found in this work (Scheme 1).

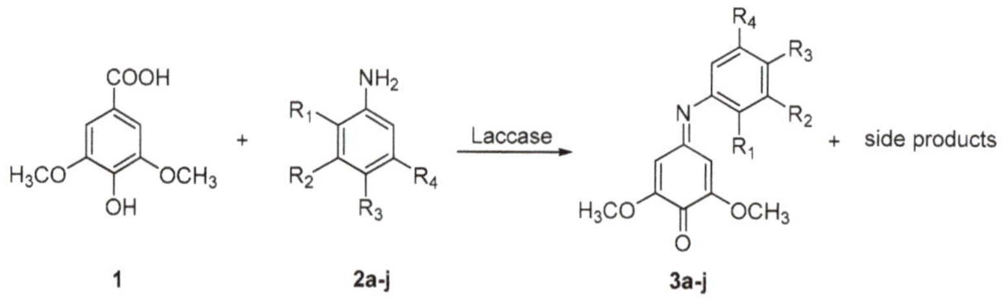

Compounds	R_1	R_2	R_3	R_4
2a/3a	H	H	Cl	H
2b/3b	H	Cl	H	H
2c/3c	H	H	OCH_3	H
2d/3d	H	OCH_3	H	H
2e/3e	H	H	NO_2	H
2f/3f	H	Cl	Cl	H
2g/3g	H	H	CF_3	H
2h/3h	H	Cl	H	Cl
2i/3i	Cl	H	H	Cl
2j/3j	Cl	H	H	H

Scheme 1. Reaction scheme for laccase-mediated synthesis between 1 and 2a–j.

To determine the reaction yields in the formation of the products, different substrate ratios were analyzed. Excluding reactions 1, 7 and 10, most reactions reached higher yields using ratio 1:1 (syringic acid:aniline) (Table 1). Moreover, when aniline was used as substrate, the same result was reported [27], indicating that the increase of the concentration of one of them decrease the yield of the obtained compounds.

Highest yields were obtained using 3-chloroaniline and 3,5-dichloroaniline as substrates (**2b** and **2h**) (Table 1). This high yield could be explained because the oxidation by laccase (from *Trametes versicolor*) of 3-chloroaniline does not occur [33]. On the other hand, using methoxyanilines (**2c** and **2d**) low yields were obtained, due to a high amount of side products (data not shown).

On the other hand, yield did not increase when the enzyme concentration was increased (data not shown). Bollag et al. [31] showed that the prolonged incubation or higher enzyme amounts caused further polymerization reaction decreasing cross-coupling formation. Furthermore, Itoh et al. [34] concluded that reactivity of laccase mediated reaction between phenolic acids and chlorophenols is due to the substrate specificity of the laccase rather than the chemical property of the substrates, which could explain the lack of relations among electron donating and withdrawing groups and yield of the reactions.

Table 1. Percentage yields of products at different reactant ratios.

Reaction	Product	Yield (%) Substrate Ratio (Syringic Acid: Aniline)		
		1:2	1:1	2:1
1	3a	10.4	36.9	55.5
2	3b	38.1	72.0	24.1
3	3c	15.5	26.7	ND [1]
4	3d	8.2	24.2	4.1
5	3e	22.6	56.8	36.4
6	3f	41.8	50.4	44.8
7	3g	12.7	10.2	8.2
8	3h	15.8	74.0	29.0
9	3i	6.9	38.3	23.9
10	3j	39.7	13.6	44.9

[1] ND: Not determined.

The ten synthetized compounds were purified using semipreparative chromatography and were identified (Figure 1) using ^{1}H-NMR and ^{13}C-NMR spectra and mass spectrometry (Figure S1–S30, Supplementary Materials). Compound **3b** showed two aliphatic proton signals (δ 3.670 (s, 3H, H8) and δ 3.874 (s, 3H, H7)) and two aliphatic carbon signals (δ 56.212 and δ 56.321) that determined the presence of two methoxy moieties. Two olefin hydrogen signals at higher fields (δ 6.010 (d, 1H, H3 J = 1.9 Hz) and δ 6.368 (d, 1H, H5 J = 1.9 Hz)), the olefin carbon signals (δ 98.583 (C3) and δ 111.717 (C5)) and one carbon signal at δ 176.633 (C1) indicated the quinonoid character of the products. Table 2 presents the NMR data (chemical shift assignments for short and long-range heteronuclear coupling) of compound **3b**.

The only difference in spectra signals between compounds **3a** and **3b** (Figures S2, S3, S5 and S6) was in the aromatic region of the spectra. Compound **3b** showed four aromatic proton signals δ 6.737 (d, 1H, H6', J = 8.0 Hz), δ 6.888 (s, 1H, H2'), δ 7.159 (d, 1H, H4', J = 8.0 Hz), and δ 7.316 (t, 1H, H5', J = 8.0 Hz) that indicated the existence of a *m*-substituted aromatic fragment. The assignation of the entire molecule was achieved by using two-dimensional NMR analysis. Therefore, identifying this compound as 4-(3'-chlorophenylimino)-2,6-dimethoxycyclohexa-2,5-dienone (Figure 1).

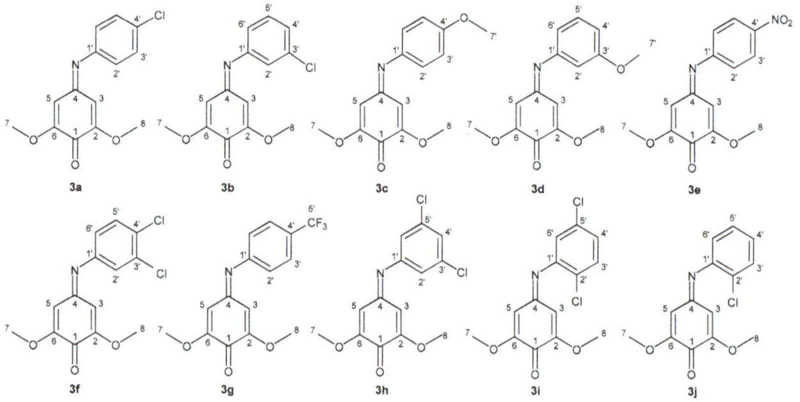

Figure 1. Structure of synthetized compounds 3a–j 2,6-dimethoxy-4-(phenylimino)cyclohexa-2,5-dienone derivatives.

Table 2. ^{1}H- and ^{13}C- chemical shifts assignments for compound 3b. Information from HSQC and HMBC experiments are also provided.

	^{13}C	^{1}H	HMBC	HSQC
1	176,633		5, 3	
4	157,816			
2	155.741		8	
6	154.881		7	
1'	151.483		5'	
3'	134.928		5'	
5'	130.247	7.316 (t, J 8.0 Hz, 1H)		5'
4'	125.023	7.159 (d, J 8.0 Hz, 1H)	6', 2'	4'
2'	120.635	6.888 (s, 1H)	6', 4'	2'
6'	118.705	6.737 (d, J 7.9 Hz, 1H)	4', 2'	6'
5	111.717	6.368 (d, J 1.9 Hz, 1H)	3	5
3	98.583	6.010 (d, J 1.9 Hz, 1H)	5	3
7	56.321	3.874 (s, 3H)		7
8	56.212	3.670 (s, 3H)		8

The spectra of compounds **3c–j** (Figures S8, S9, S11, S12, S14, S15, S17, S18, S20, S21, S23, S24, S26, S27, S29 and S30) only showed differences in the aromatic region; the assignment of the ^{1}H and ^{13}C-NMR spectra can be found in the spectroscopic data section (Section 3.4). Figure 1 shows the structures of the ten synthetized compounds in this work. To our knowledge compounds **3a** and **3f** were previously synthesized [31], ^{1}H-NMR spectra for compounds **3a** and **3f** (spectroscopic data Section 3.4) have the same number of signals and comparable chemical shifts and coupling constants like those found by Bollag et al. [31]; furthermore, the mass spectra of **3a** and **3f** showed a base peak with m/z 277 and 311, respectively, corresponding to the molecular ions, equivalent to the previously described data [31]. Therefore, the other eight compounds (**3b**, **3c**, **3d**, **3e**, **3g**, **3h**, **3i** and **3j**) have not been previously reported.

Interesting, compound **3a** was also obtained using syringaldehyde instead of syringic acid in the reaction with 4-chloroaniline. This could be explained with an extra step when using syringaldehyde, an oxidation of the aldehyde to a carboxylic acid (syringic acid), similar oxidations has been previously described using several aromatic aldehydes with laccase, yielding carboxylic acids [35]. Hence, syringaldehyde is oxidized to syringic acid and then the same product (compound **3a**) could be found in both reactions, starting with syringaldehyde or with syringic acid. However, this synthesis had a very low yield (data not shown).

2.2. Antifungal Activity

Antifungal activity of compounds **1** and **2a–j** and compounds **3a–j** against *B. cinerea* was measured on mycelial growth in solid media and the EC_{50} were calculated using the mycelial growth (Tables 3 and 4). The most active compound was the 3,5-dichloro-substituted product (compound **3h**), while compound **3c** had the lowest activity. It has been reported that the substituent affects the antifungal activity of a molecule [36], for instance, the position of the chlorine atom in the aromatic ring is important for the antifungal activity against *B. cinerea* since para-substituted compound (**3a**) and ortho-substituted compound (**3j**) were more active than the meta-substituted compound and unsubstituted compound (EC_{50} = 0.14 ± 0.02) [27], while activity of meta-substituted compound (compound **3b**) and unsubstituted compound are similar [27]. The number of chlorine atoms in the aromatic ring is also important, both dichlorinated compounds **3f** and **3h** showed higher antifungal activity than mono chlorinated compounds **3a**, **3b**, and **3j**, however, dichlorinated compound **3i** showed an antifungal activity comparable to the monochlorinated compounds, therefore, the number and position of chlorine atoms in the aromatic ring seems to be important for the antifungal activity of these compounds.

Table 3. Effect of compounds **3a–j** on the mycelial growth of *B. cinerea* in solid medium.

Compound	EC_{50} (mM)
3a	0.065 ± 0.003
3b	0.15 ± 0.01
3c	0.54 ± 0.06
3d	0.39 ± 0.02
3e	- *
3f	0.055 ± 0.004
3g	0.101 ± 0.014
3h	0.032 ± 0.003
3i	0.065 ± 0.011
3j	0.069 ± 0.004

* No activity.

Table 4. Effect of the substrates on the mycelial growth of *B. cinerea* in solid medium.

Compound	EC_{50} (mM)
1	>1.51
2a	0.71 ± 0.08
2b	0.59 ± 0.04
2c	>3.00
2d	>3.00
2e	0.15 ± 0.01
2f	0.047 ± 0.005
2g	2.058 ± 0.434
2h	0.12 ± 0.02
2i	0.414 ± 0.031
2j	1.09 ± 0.08

Furthermore, the methoxy derivative compounds (**3c** and **3d**) were less active against the fungus than the other compounds, even the nonsubstituted compound **3** [27], the same effect was observed for aspirin derivatives, where the methoxy para-substituted derivative showed almost 30% less antifungal activity against *B. cinerea* than the chlorinated para-substituted compound [37]. Similar behavior was previously reported for oxadiazole derivatives when tested the activity of the methoxy meta-substituted oxadiazole derivative against *B. cinerea*, and its activity was less than half compared to the nonsubstituted compound [38]. Usually, the chloro-substituted compounds have higher antifungal activity in commercial fungicides, for example, chlorine compounds like boscalid, chlorothalonil, and iprodione have been used to control *B. cinerea* [6]. The antifungal activity of iprodione has been tested against this strain of *B. cinerea*, showing an EC_{50} of 0.015 ± 0.003 mM [27], this antifungal activity is in the same order of magnitude than the most active compound obtained in this work (**3h**). Additionally, *p*-nitro and *p*-trifluoromethyl compounds (**3e** and **3g**) were tested against this fungus, **3e** showed no antifungal activity, probably because of the low solubility of this molecule, for this reason **3e** was not used in further assays. Compound **3g** only showed an intermediate antifungal activity compared to the rest of the synthetized molecules in this work. Lastly, most of the substrates used in the reactions (i.e., **1** and **2a–j**) showed lower antifungal activity than the products (Table 4), only **2f** was more active than **3f**.

2.3. Effect on the Cell Wall Integrity of B. cinerea

To analyze the effect of the compounds on the cell wall integrity, the dye calcofluor white (CFW) was used. This dye binds to β-1,3 and β-1,4 polysaccharides, for example chitin, which is a primary component of the cell wall in fungi, and fluorescence of the hyphae can be detected [39]. Figure 2 shows the effect of compound **3a** on the cell wall of *B. cinerea*. Treatment with this compound showed lower fluorescence intensity than the negative control (acetone), indicating that this compound can damage the cell wall of this fungus. The same assay was performed using compounds **3b–j**. The ten synthesized compounds caused a decrease of the fluorescence intensity compared to the control; relative fluorescence intensity is observed in Figure 3. This result could be attributed to the toxicity of quinones, which could be connected to the production of reactive oxygen species (ROS) which cause oxidation of cell molecules [40]. Quinone derivative N-acetyl-*p*-benzoquinone imine (NAPQI) can react with nucleophiles such as thiol groups of proteins or glutathione [41,42]; this last molecule is an important antioxidant molecule in fungi [43]. On the other hand, some aromatic antifungal compounds have shown effects on cell wall synthesis [6] by inhibiting chitin and glucan synthases [44], enzymes that catalyze the synthesis of the main polymers of the cell wall in fungi.

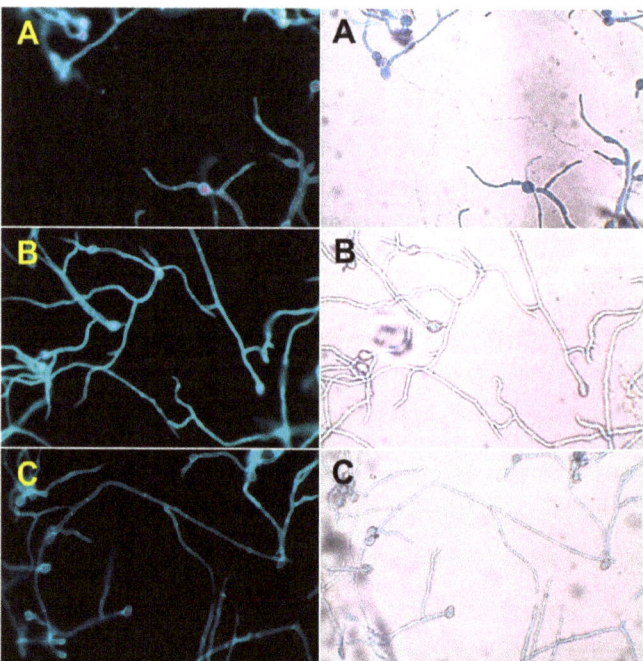

Figure 2. Effect of compound **3a** on the cell wall of *B. cinerea*. Hyphae of *B. cinerea* incubated with liquid medium along with (**A**) lysing enzymes (positive control), (**B**) acetone 5% (*v/v*) (negative control), and (**C**) compound **3a** at 0.16 mM. *B. cinerea* hyphae were treated with calcofluor white (CFW) stain. Assays were carried out in triplicate.

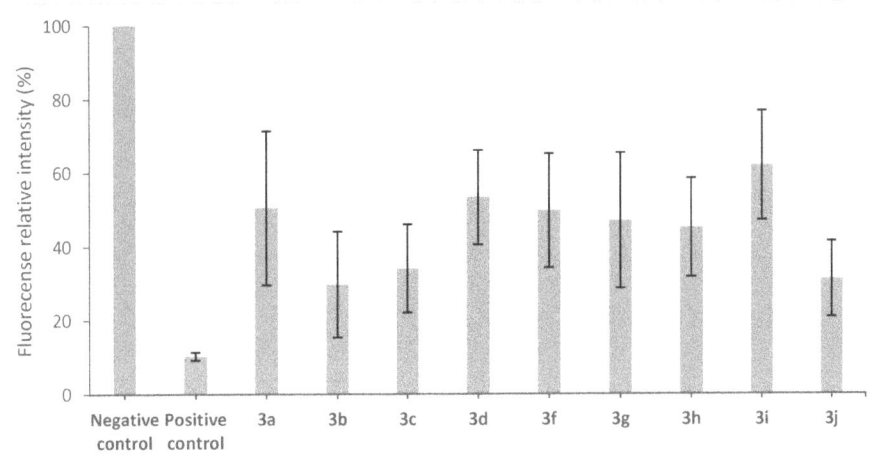

Figure 3. Effect of the compounds on the cell wall integrity of *B. cinerea*. The effect was measured as relative fluorescence intensity compared to maximum fluorescence (negative control).

3. Materials and Methods

3.1. General Experimental Procedures

The NMR spectra of **3a–j** were acquired using a Bruker Avance 400 MHz spectrometer (Bruker, Billerica, MA, USA) (400,133 MHz for ^{1}H, 100.624 MHz for ^{13}C). Measurements were done in CDCl$_3$ at 27 °C. Chemical shifts were calibrated to solvent signal: CHCl$_3$ 7.26 ppm (residual signal solvent) and 77.16 ppm for ^{1}H and ^{13}C, respectively, and informed relative to Me$_4$Si. Thin-layer chromatography was done with a Merck Kiesegel 60 F$_{254}$, 0.2 mm thick and semipreparative thin layer chromatography on Merck Kieselgel 60 F$_{254}$ 0.25 mm thick. A Thermo Scientific GC-MS system (GC: model: Trace 1300 and MS: model TSQ8000Evo) (Waltham, Massachusetts, USA) was used to analyze the sample. The separation was performed on a 60 m × 0.25 mm internal diameter fused silica capillary column coated with 0.25 μm film Rtx-5MS. The oven temperature was maintained at 40 °C for 5 min, then it was programmed from 40 to 80 at 5 °C/min for 1 min, then from 80 to 300 at 10 °C/min and finally maintained at 300 °C for 10 min. The mode used was splitless injection, helium was used as carrier gas, and flow-rate was 1.2 mL/min. Mass spectra were recorded over a range of 40 to 400 atomic mass units at 0.2 s/scan. Solvent cut time was 11 min. Ionization energy was 70 eV.

3.2. Chemical Reagents

Laccase from *Trametes versicolor* (EC 1.10.3.2), lysing enzymes from *Trichoderma harzianum*, Calcofluor white stain, 4-hydroxy-3,5-dimethoxy-benzoic acid (syringic acid), 3,5-dimethoxy-4-hydroxybenzaldehyde (syringaldehyde), 4-chloroaniline, 2,5-dichloroaniline, 3,5-dichloroaniline, and 4-nitroaniline were obtained from Sigma Chemical Co. (St. Louis, MO, USA). 3-chloroaniline, 4-methoxyaniline, 3-methoxyaniline, 3,4-dichloroaniline, 2-chloroaniline, organic solvents, and salts were obtained from Merck (Hohenbrunn, Germany). 4-(trifluoromethyl)aniline was obtained from Santa Cruz Biotechnology (Finnell St, Dallas Tx). Agar was obtained from Difco Laboratories (Detroit, MI, USA).

3.3. Laccase-Mediated Synthesis of 2,6-Dimethoxy-4-(phenylimino)cyclohexa-2,5-dienone Derivatives (Compounds 3a–j)

Syringic acid with an aniline derivative at different ratios (1:1, 1:2, and 2:1) (e.g., for ratio 1:1 means 0.1 mmol for both syringic acid and the aniline derivative were used) and different enzyme quantities (2.25, 4.5, and 9 U) were tested to increase the yield of the synthesized compounds.

For the first reaction, syringic acid (1) and 4-chloroaniline (2a) were dissolved in 1 mL ethyl acetate and laccase was dissolved in 1 mL sodium acetate buffer (20 mM, pH 4.5). Both solutions were mixed and stirred at 100 rpm for 180 min at 22 °C. Afterwards, the solvent was evaporated at 40 °C using a rotary evaporator. The synthetized compounds were purified by using semipreparative thin layer chromatography with hexane: ethyl acetate (1:1) as an eluent system. Same procedure was carried out using a different substituted aniline (2b–j).

Alternatively, compound 3a was also found when using syringaldehyde and 4-chloroaniline under the same conditions described above.

3.4. Spectroscopic Data

Compound 3a (4-(4′-chlorophenylimino)-2,6-dimethoxycyclohexa-2,5-dienone) Yield 55.5%. ^{1}H-NMR (CDCl$_3$, 400 MHz) δ 3.668 (s, 3H, H8), 3.870 (s, 3H, H7), 6.040 (d, 1H, J = 1.9 Hz, H3), 6.377 (d, 1H, J = 1.9 Hz, H5), 6.816 (d, 2H, J = 8.5 Hz, H2′), 7.361(d, 2H, J = 8.5 Hz, H3′); ^{13}C-NMR (CDCl$_3$, 100 MHz) δ 56.188 (C8), 56.298 (C7), 98.529 (C3), 111.846 (C5), 122.040 (C2′), 129.333 (C3′), 130.749 (C4′), 148.777 (C1′), 154.813 (C6), 155.745 (C2), 157.642 (C4), 176.647 (C1). mp 208.0–209.1 °C. GC-MS RI$_{(Rtx-5ms)}$ = 2366, C$_{14}$H$_{12}$O$_3$NCl EI-MS m/z: 111 (15); 150 (16); 178 (15); 182 (17); 197 (35); 212 (15); 224 (22); [M]$^+$ = 277 (100); [M + 1]$^+$ = 278 (17); [M + 2]$^+$ = 279 (36).

Compound 3b (4-(3′-chlorophenylimino)-2,6-dimethoxycyclohexa-2,5-dienone) Yield 72.0%. ^{1}H-NMR (CDCl$_3$, 400 MHz) δ 3.670 (s, 3H, H8), 3.874 (s, 3H, H7), 6.010 (d, 1H, J = 1.9 Hz, H3), 6.368 (d, 1H, J = 1.9 Hz, H5), 6.737 (d, 1H, J = 7.9 Hz, H6′), 6.888 (s, 1H, J = 8.0 Hz, H2′), 7.159 (d, 1H, J = 8,0 Hz, H4′), 7.316 (t, 1H, J = 7.9 Hz, H5′); ^{13}C-NMR (CDCl$_3$, 100 MHz) δ 56.212 (C8), 56.321 (C7), 98.583 (C3), 111.717 (C5), 118.705 (C6′), 120.635 (C2′), 125.023 (C4′), 130.247 (C5′), 134.928 (C3′), 151.483 (C1′), 154.881 (C6), 155.741 (C2), 157.816 (C4), 176.633 (C1). mp 155.7–156.0 °C. GC-MS RI$_{(Rtx-5ms)}$ = 2340, C$_{14}$H$_{12}$O$_3$NCl EI-MS m/z: 69 (22); 75 (52); 111 (58); 113 (21); 140 (20); 178 (26); 182 (24); 197 (43) [M]$^+$ = 277 (100); [M + 1]$^+$ = 278 (17); [M + 2]$^+$ =279 (36).

Compound 3c (4-(4′-methoxyphenylimino)-2,6-dimethoxycyclohexa-2,5-dienone) Yield 23.7%. ^{1}H-NMR (CDCl$_3$, 400 MHz) δ 3.691 (s, 3H, H8), 3.846 (s, 3H, H7′), 3.870 (s, 3H, H7), 6.242 (d, 1H, J = 1.9 Hz, H3), 6.467 (d, 1H, J = 1.9 Hz, H5), 6.903 (d, 2H, J = 8.9 Hz, H2′), 6.956 (d, 2H, J = 8.9 Hz, H3′); ^{13}C-NMR (CDCl$_3$, 100 MHz) δ 55.656 (C7′), 56.104 (C8), 56.277 (C7), 98.966 (C3), 112.113 (C5), 114.594 (C3′), 123.047 (C2′), 143.196 (C1′), 154.661 (C6), 155.739 (C2), 156.809 (C4), 158.018 (C4′), 176.669 (C1). mp 111.7–112.3 °C. GC-MS RI$_{(Rtx-5ms)}$ = 2451, C$_{15}$H$_{15}$O$_4$N EI-MS m/z: 134 (10); 172 (20); 198 (12); 200 (9); 212 (9); 230 (30); 240 (12); 258 (45); [M]$^+$ = 273 (100); [M + 1]$^+$ = 274 (18).

Compound 3d (2,6-dimethoxy-4-(3′-methoxyphenylimino)cyclohexa-2,5-dienone) Yield 24.24%. ^{1}H-NMR (CDCl$_3$, 400 MHz) δ 3.656 (s, 3H, H8), 3.819 (s, 3H, H7′), 3.868 (s, 3H, H7), 6.125 (d, 1H, J = 2.0 Hz, H3), 6.392 (d, 1H, J = 2.0 Hz, H5), 6.429 (d, 1H, J = 7.7 Hz, H6′), 6.452 (s, 1H, H2′), 6.736 (d, 1H, J = 8.1 Hz, H4′), 7.281 (t, 1H, J = 8.0 Hz, H5′); ^{13}C-NMR (CDCl$_3$, 100 MHz) δ 55.476 (C7′), 56.124 (C8), 56.248 (C7), 99.051 (C3), 106.387 (C2′), 110.942 (C4′), 112.023 (C5), 112.868 (C6′), 126.946 (C5′), 151.664 (C1′), 154.783 (C6), 155.569 (C2), 157.264 (C4), 160.402 (C3′), 176.785 (C1). mp 133.5–134.6 °C. GC-MS RI$_{(Rtx-5ms)}$ = 2403, C$_{15}$H$_{15}$O$_4$N EI-MS m/z: 159 (13); 187 (16); 199 (19); 200 (13); 212 (12); 215 (22); 230 (21); 242 (24) [M]$^+$ = 273 (100); [M + 1]$^+$ = 274 (18).

Compound **3e** (*2,6-dimethoxy-4-(4′-nitrophenylimino)cyclohexa-2,5-dienone*) Yield 56.8%. ^{1}H-NMR (CDCl$_3$, 400 MHz) δ 3.660 (s, 3H, H8), 3.895 (s, 3H, H7), 5.845 (d, 1H, J = 1.8 Hz, H3), 6.363 (d, 1H, J = 1.8 Hz, H5), 6.959 (d, 2H, J = 8.8 Hz, H2′), 8.279(d, 2H, J = 8.8 Hz, H3′); ^{13}C-NMR (CDCl$_3$, 100 MHz) δ 56.341 (C8), 56.439 (C7), 98.339 (C3), 111.172 (C5), 120.620 (C2′), 125.205 (C3′), 144.864 (C4′), 155.257 (C1′), 156.102 (C6), 156.170 (C2), 157.940 (C4), 176.321 (C1). mp 208.0–209.1 °C. GC-MS RI$_{(Rtx-5ms)}$ = 2690, C$_{14}$H$_{12}$O$_5$N$_2$ EI-MS m/z: 16 (33); 128 (21); 143 (29); 156 (25); 168 (18); 169 (20); 197 (38); 211 (26); [M]$^+$ = 288 (100); [M + 1]$^+$ = 289 (16).

Compound **3f** (*4-(3′,4′-dichlorophenylimino)-2,6-dimethoxycyclohexa-2,5-dienone*) Yield 50.4%. ^{1}H-NMR (CDCl$_3$, 400 MHz) δ 3.690 (s, 3H, H8), 3.874 (s, 3H, H7), 5.982 (s, 1H, H3), 6.347 (s, 1H, H5), 6.715 (d, 1H, J = 8.4 Hz, H6′), 6.994 (s, 1H, H2′), 7.449 (d, 1H, J = 8.4 Hz, H5′); ^{13}C-NMR (CDCl$_3$, 100 MHz) δ 56.319 (C8), 56.355 (C7), 98.322 (C3), 111.572 (C5), 120.097 (C6′), 122.367 (C2′), 128.704 (C3′), 130.913 (C5′), 133.211 (C4′), 149.770 (C1′), 155.033 (C6), 155.973 (C2), 158.242 (C4), 176.486 (C1). mp 160.1–161.5 °C. GC-MS RI$_{(Rtx-5ms)}$ = 2556, C$_{14}$H$_{11}$O$_3$NCl$_2$ EI-MS m/z: 109 (22); 145 (18); 184 (21); 212 (20); 216 (26); 231 (41); 233 (28); 258 (24); [M]$^+$ = 311 (100); [M + 2]$^+$ = 313 (65); [M + 4]$^+$ = 315 (13).

Compound **3g** (*4-(4′-(trifluoromethyl)phenylimino)-2,6-dimethoxycyclohexa-2,5-dienone*) Yield 12.7%. ^{1}H-NMR (CDCl$_3$, 400 MHz) δ 3.643 (s, 3H, H8), 3.868 (s, 3H, H7), 5.928 (s, 1H, H3), 6.367 (s, 1H, H5), 6.933 (d, 2H, J = 8.1 Hz, H2′), 7.633 (d, 2H, J = 8.1 Hz, H3′); ^{13}C-NMR (CDCl$_3$, 100 MHz) δ 56.211 (C8), 56.302 (C7), 98.439 (C3), 111.559 (C5), 120.453 (C2′), 124.319 (q, J = 271.6 Hz, C5′), 126.425 (q, J = 3.7 Hz, C3′), 126.936 (q, J = 32.7 Hz, C4′), 153.272 (C1′), 154.987 (C6), 155.898 (C2), 157.775 (C4), 176.491 (C1). mp 130–133 °C. GC-MS RI$_{(Rtx-5ms)}$ = 2084, C$_{15}$H$_{12}$O$_3$NF$_3$ EI-MS m/z: 53 (22); 69 (43); 95 (37); 125 (30); 145 (96); 184 (29); 197 (57); 212 (52); 221 (22); [M]$^+$ = 311 (100).

Compound **3h** (*4-(3′,5′-dichlorophenylimino)-2,6-dimethoxycyclohexa-2,5-dienone*) Yield 74.0%. ^{1}H-NMR (CDCl$_3$, 400 MHz) δ 3.699 (s, 3H, H8), 3.874 (s, 3H, H7), 5.935 (d, 1H, J = 2.1 Hz, H3), 6.327 (d, 1H, J = 2.1 Hz, H5), 6.758 (d, 2H, J = 1.8 Hz, H2′), 7.167 (d, 1H, J = 1.8 Hz, H4′); ^{13}C-NMR (CDCl$_3$, 100 MHz) δ 56.375 (C8 and C7), 98.324 (C3), 111.375 (C5), 118.863 (C2′), 124.712 (C4′), 135.532 (C3′ and C5′), 152.147 (C1′), 155.044 (C6), 155.938 (C2), 158.388 (C4), 176.473 (C1). mp. 170–173 °C. GC-MS RI$_{(Rtx-5ms)}$ = 2472, C$_{14}$H$_{11}$O$_3$NCl$_2$ EI-MS m/z: 109 (18); 145 (19); 212 (18); 216 (19); 231 (29); 233 (22); 246 (14); [M]$^+$ = 311 (100); [M + 2]$^+$ = 313 (66); [M + 4]$^+$ = 315 (13).

Compound **3i** (*4-(2′,5′-dichlorophenylimino)-2,6-dimethoxycyclohexa-2,5-dienone*) Yield 38.3%. ^{1}H-NMR (CDCl$_3$, 400 MHz) δ 3.678 (s, 3H, H8), 3.893 (s, 3H, H7), 5.794 (d, 1H, J = 1.8 Hz, H3), 6.416 (d, 1H, J = 1.8 Hz, H5), 6.838 (d, 1H, J = 2.2 Hz, H3′), 7.094 (dd, 1H, J = 8.6, 2.2 Hz, H5′), 7.382 (d, 1H, J = 8.6 Hz, H6′); ^{13}C-NMR (CDCl$_3$, 100 MHz) δ 56.418 (C8 and C7), 98.759 (C3), 111.205 (C5), 121.133 (C3′), 123.120 (C1′), 125.644 (C5′), 131.130 (C6′), 133.044 (C4′), 148.282 (C2′), 155.162 (C6), 155.840 (C2), 159.208 (C4), 176.453 (C1). Mp. 198–202 °C. GC-MS RI$_{(Rtx-5ms)}$ = 2449, C$_{14}$H$_{11}$O$_3$NCl$_2$ EI-MS m/z: 190 (24); 212 (25); 231 (29); 233 (79); 261 (25); 276 (92); 278 (29); [M]$^+$ = 311 (100); [M + 2]$^+$ = 313 (73); [M + 4]$^+$ = 315 (15).

Compound **3j** (*4-(2′-chlorophenylimino)-2,6-dimethoxycyclohexa-2,5-dienone*) Yield 44.9%. ^{1}H-NMR (CDCl$_3$, 400 MHz) δ 3.637 (s, 3H, H8), 3.887 (s, 3H, H7), 5.862 (s, 1H, H3), 6.460 (s, 1H, H5), 6.797 (d, 1H, J = 7.7 Hz, H6′), 7.119 (t, 1H, J = 7.6 Hz, H5′), 7.274 (t, 1H, J = 7.6 Hz, H4′), 7.458 (d, 1H, J = 8.0 Hz, H3′); ^{13}C-NMR (CDCl$_3$, 100 MHz) δ 56.158 (C8), 56.332 (C7), 99.002 (C3), 111.548 (C5), 121.317 (C6′), 124.984 (C1′), 125.949 (C5′), 127.326 (C4′), 130.277 (C3′), 147.388 (C2′), 154.971 (C6), 155.618 (C2), 158.576 (C4), 176.616 (C1). mp 143–146 °C. GC-MS RI$_{(Rtx-5ms)}$ = 2449, C$_{14}$H$_{12}$O$_3$NCl EI-MS m/z: 150 (17); 170 (16); 178 (21); 197 (25); 199 (54); 214 (14); 242 (54); [M]$^+$ = 277 (100); [M + 1]$^+$ = 278 (16); [M + 2]$^+$ = 279 (36).

3.5. Fungal Strain and Culture Conditions

The strain G29 of *B. cinerea* used in this work was isolated from infected grapes (*Vitis vinifera*) and genetically characterized by the INIA, La Platina, Chile [45]. It was kept on malt yeast extract agar slants with 0.2% (w/v) yeast extract, 2% (w/v) malt extract, and 1.5% (w/v) agar) at 4 °C. For the cell wall integrity assay, liquid minimal medium of pH 6.5 was used, containing KH$_2$PO$_4$ (1 g/L), MgSO$_4$·7H$_2$O

(0.5 g/L), KCl (0.5 g/L), K$_2$HPO$_4$ (0.5 g/L), FeSO$_4$·7H$_2$O (0.01 g/L), 1% (w/v) glucose as a carbon source, and 4.6 g/L ammonium tartrate as a nitrogen source.

3.6. Antifungal Assay

Effect on Mycelial Growth

The antifungal activity of the compounds was evaluated in vitro as described by Caruso et al. [13]. Compounds were dissolved in acetone and then added to Petri dishes along with malt yeast agar medium. Inhibition percentages were calculated after 72 h of incubation. Antifungal activity was expressed as the concentration that reduced mycelial growth by 50% (EC$_{50}$), calculated by regressing the antifungal activity percentage against compound concentration. These experiments were done at least in triplicate.

3.7. Effect on the Cell Wall Integrity of B. cinerea

The effect of compounds **3a–j** on cell wall integrity was evaluated using the method described by Mendoza et al. [27]. Compounds **3a–j** were tested at 0.16 mM. To measure the effect of these compounds on the cell wall, fluorescence intensity was quantified using ImageJ (v1.80), an outline was drawn around each hypha, and mean fluorescence was measured, along with several adjacent background readings. Mean fluorescence was compared to the negative control (maximum fluorescence).

4. Conclusions

Ten compounds were synthetized; two of them (compounds **3a** and **3f**) have been previously described. All the products showed higher antifungal activity than the substrates. Chloro-substituted compounds showed the highest antifungal effect against *B. cinerea* being the 3,5-dichlorinated product **3h** the most active. Synthesis using syringic acid or syringaldehyde with *p*-chloroaniline yield the same main product (compound **3a**). Finally, regarding the inhibition mechanism of these compounds, the results suggest that these compounds damage the cell wall.

Supplementary Materials: The following are available online at http://www.mdpi.com/1420-3049/24/4/706/s1, Figures S1–S30: ^{1}H-, ^{13}C-NMR and Mass spectra of compounds **3a–j**, Figures S31–S40: IR spectra of compounds **3a–j**. Table S1: Retention Index.

Author Contributions: Methodology, P.C., P.C.P., K.L., and R.S.; Investigation, P.C.; Formal Analysis, L.M., M.C., F.N. and P.C.; Writing-Original Draft P.C.; Writing-Review and Editing, L.M., M.C., F.N., P.C.P., and P.C.; Project Administration C.V.

Funding: This research was funded by FONDECYT, grant number 3170478, and by Universidad de Santiago de Chile DICYT 021640VG_POSTDOC.

Acknowledgments: CONICYT FONDEQUIP GC-MS/MS EQM 150084.

Conflicts of Interest: The authors declare no conflict of interest. The funders had no role in the design of the study; in the collection, analyses, or interpretation of data; in the writing of the manuscript, or in the decision to publish the results.

References

1. Fillinger, S.; Elad, Y. *Botrytis—The Fungus, the Pathogen and Its Management in Agricultural Systems*; Springer: Cham, Switzerland, 2016; pp. 189–216.
2. Fenner, K.; Canonica, S.; Wackett, L.P.; Elsner, M. Evaluating pesticide degradation in the environment: Blind spots and emerging opportunities. *Science* **2013**, *341*, 752–758. [CrossRef] [PubMed]
3. Brent, K.J.; Hollomon, D.W. *Fungicide Resistance: The Assessment of the Risk*; Fungicide Resistance Action Committee: Brussels, Belgium, 2007; pp. 1–53.
4. Pappas, A.C. Evolution of fungicide resistance in *Botrytis cinerea* in protected crops in Greece. *Crop Prot.* **1997**, *16*, 257–263. [CrossRef]

5. Wilson, C.L.; Solar, J.M.; Ghaouth, A.E.; Wisniewski, M.E. Rapid evaluation of plant extracts and essential oils for antifungal activity against *Botrytis cinerea*. *Plant Dis.* **1997**, *81*, 204–210. [CrossRef]
6. Elad, Y.; Williamson, B.; Tudzynski, P.; Delen, N. *Botrytis: Biology, Pathology and Control*; Springer: Dordrecht, The Netherlands, 2007; pp. 195–217.
7. Daoubi, M.; Durán-Patrón, R.; Hmamouchi, M.; Hernández-Galán, R.; Benharref, A.; Collado, I.G. Screening study for potential lead compounds for natural product-based fungicides: I. Synthesis and in vitro evaluation of coumarins against *Botrytis cinerea*. *Pest Manag. Sci.* **2004**, *60*, 927–932. [CrossRef] [PubMed]
8. Imperato, F. *Phytochemistry: Advances in Research*; Research Signpost: Trivandrum, India, 2006; pp. 23–67.
9. Grayer, R.J.; Kokubun, T. Plant–fungal interactions: The search for phytoalexins and other antifungal compounds from higher plants. *Phytochemistry* **2001**, *56*, 253–263. [CrossRef]
10. Grayer, R.J.; Harborne, J.B. A survey of antifungal compounds from higher plants, 1982–1993. *Phytochemistry* **1994**, *37*, 19–42. [CrossRef]
11. Osbourn, A.E. Preformed antimicrobial compounds and plant defense against fungal attack. *Plant Cell* **1996**, *8*, 1821–1831. [CrossRef]
12. Harborne, J.B. The comparative biochemistry of phytoalexin induction in plants. *Biochem. Syst. Ecol.* **1999**, *27*, 335–367. [CrossRef]
13. Caruso, F.; Mendoza, L.; Castro, P.; Cotoras, M.; Aguirre, M.; Matsuhiro, B.; Isaacs, M.; Rossi, M.; Viglianti, A.; Antonioletti, R. Antifungal activity of resveratrol against *Botrytis cinerea* is improved using 2-furyl derivatives. *PLoS ONE* **2011**, *6*, e25421. [CrossRef]
14. Pinedo-Rivilla, C.; Bustillo, A.J.; Hernández-Galán, R.; Aleu, J.; Collado, I.G. Asymmetric preparation of antifungal 1-(4′-chlorophenyl)-1-cyclopropyl methanol and 1-(4′-chlorophenyl)-2-phenylethanol. Study of the detoxification mechanism by *Botrytis cinerea*. *J. Mol. Catal. B Enzym.* **2011**, *70*, 61–66. [CrossRef]
15. Saiz-Urra, L.; Racero, J.C.; Macías-Sáchez, A.J.; Hernández-Galán, R.; Hanson, J.R.; Perez-Gonzalez, M.; Collado, I.G. Synthesis and quantitative structure-antifungal activity relationships of clovane derivatives against *Botrytis cinerea*. *J. Agric. Food Chem.* **2009**, *57*, 2420–2428. [CrossRef] [PubMed]
16. Mendoza, L.; Yañez, K.; Vivanco, M.; Melo, R.; Cotoras, M. Characterization of extracts from winery by-products with antifungal activity against *Botrytis cinerea*. *Ind. Crops Prod.* **2013**, *43*, 360–364. [CrossRef]
17. Mikolasch, A.; Schauer, F. Fungal laccases as tools for the synthesis of new hybrid molecules and biomaterials. *Appl. Microbiol. Biotechnol.* **2009**, *82*, 605–624. [CrossRef] [PubMed]
18. Senthivelan, T.; Kanagaraj, J.; Panda, R.C. Recent trends in fungal laccase for various industrial applications: An eco-friendly approach—A review. *Biotechnol. Bioprocess Eng.* **2016**, *21*, 19–38. [CrossRef]
19. Mogharabi, M.; Faramarzi, M.A. Laccase and laccase-mediated systems in the synthesis of organic compounds. *Adv. Synth. Catal.* **2014**, *356*, 897–927. [CrossRef]
20. Constantin, M.-A.; Conrad, J.; Merişor, E.; Koschorreck, K.; Urlacher, V.B.; Beifuss, U. Oxidative dimerization of (*E*)- and (*Z*)-2-propenylsesamol with O_2 in the presence and absence of laccases and other catalysts: Selective formation of carpanones and benzopyrans under different reaction conditions. *J. Org. Chem.* **2012**, *77*, 4528–4543. [CrossRef] [PubMed]
21. Koschorreck, K.; Richter, S.M.; Ene, A.B.; Roduner, E.; Schmid, R.D.; Urlacher, V.B. Cloning and characterization of a new laccase from *Bacillus licheniformis* catalyzing dimerization of phenolic acids. *Appl. Microbiol. Biotechnol.* **2008**, *79*, 217–224. [CrossRef]
22. Adelakun, O.E.; Kudanga, T.; Green, I.R.; le Roes-Hill, M.; Burton, S.G. Enzymatic modification of 2,6-dimethoxyphenol for the synthesis of dimers with high antioxidant capacity. *Process Biochem.* **2012**, *47*, 1926–1932. [CrossRef]
23. Adelakun, O.E.; Kudanga, T.; Parker, A.; Green, I.R.; le Roes-Hill, M.; Burton, S.G. Laccase-catalyzed dimerization of ferulic acid amplifies antioxidant activity. *J. Mol. Catal. B Enzym.* **2012**, *74*, 29–35. [CrossRef]
24. Navarra, C.; Gavezzotti, P.; Monti, D.; Panzeri, W.; Riva, S. Biocatalyzed synthesis of enantiomerically enriched β-5-like dimer of 4-vinylphenol. *J. Mol. Catal. B Enzym.* **2012**, *84*, 115–120. [CrossRef]
25. Mikolasch, A.; Niedermeyer, T.H.J.; Lalk, M.; Witt, S.; Seefeldt, S.; Hammer, E.; Schauer, F.; Gesell Salazar, M.; Hessel, S.; Jülich, W.-D.; et al. Novel cephalosporins synthesized by amination of 2,5-dihydroxybenzoic acid derivatives using fungal laccases II. *Chem. Pharm. Bull.* **2007**, *55*, 412–416. [CrossRef] [PubMed]
26. Mikolasch, A.; Hessel, S.; Salazar, M.G.; Neumann, H.; Manda, K.; Gordes, D.; Schmidt, E.; Thurow, K.; Hammer, E.; Lindequist, U.; et al. Synthesis of new *N*-analogous corollosporine derivatives with antibacterial activity by laccase-catalyzed amination. *Chem. Pharm. Bull.* **2008**, *56*, 781–786. [CrossRef] [PubMed]

27. Mendoza, L.; Castro, P.; Melo, R.; Campos, A.M.; Zuñiga, G.; Guerrero, J.; Cotoras, M. Improvement of the antifungal activity against *Botrytis cinerea* of syringic acid, a phenolic acid from grape pomace. *J. Chil. Chem. Soc.* **2016**, *61*, 3039–3042. [CrossRef]
28. Leroux, P. Recent developments in the mode of action of fungicides. *Pestic. Sci.* **1996**, *47*, 191–197. [CrossRef]
29. Guo, Z.; Miyoshi, H.; Komyoji, T.; Haga, T.; Fujita, T. Uncoupling activity of a newly developed fungicide, fluazinam [3-chloro-*N*-(3-chloro-2,6-dinitro-4-trifluoromethylphenyl)-5-trifluoromethyl-2-pyridinamine]. *BBA Bioenerg.* **1991**, *1056*, 89–92. [CrossRef]
30. Leroux, P.; Gredt, M.; Leroch, M.; Walker, A.S. Exploring mechanisms of resistance to respiratory inhibitors in field strains of *Botrytis cinerea*, the causal agent of gray mold. *Appl. Environ. Microbiol.* **2010**, *76*, 6615–6630. [CrossRef] [PubMed]
31. Bollag, J.M.; Minard, R.D.; Liu, S.Y. Cross-linkage between anilines and phenolic humus constituents. *Environ. Sci. Technol.* **1983**, *17*, 72–80. [CrossRef] [PubMed]
32. Tatsumi, K.; Freyer, A.; Minard, R.D. Enzymatic coupling of chloroanilines, syringic acid, vanillic acid and protocatechuic acid. *Soil Biol. Biochem.* **1994**, *26*, 135–142. [CrossRef]
33. Hoff, T.; Liu, S.Y.; Bollag, J.M. Transformation of halogen-, alkyl-, and alkoxy-substituted anilines by a laccase of *Trametes versicolor*. *Appl. Environ. Microbiol.* **1985**, *49*, 1040–1045.
34. Itoh, K.; Fujita, M.; Kumano, K.; Suyama, K.; Yamamoto, H. Phenolic acids affect transformations of chlorophenols by a *Coriolus versicolor* laccase. *Soil Biol. Biochem.* **2000**, *32*, 85–91. [CrossRef]
35. Galletti, P.; Pori, M.; Funiciello, F.; Soldati, R.; Ballardini, A.; Giacomini, D. Laccase-mediator system for alcohol oxidation to carbonyls or carboxylic acids: Toward a sustainable synthesis of profens. *ChemSusChem* **2014**, *7*, 2684–2689. [CrossRef] [PubMed]
36. Saiz-Urra, L.; Bustillo-Pérez, A.J.; Cruz-Monteagudo, M.; Pinedo-Rivilla, C.; Aleu, J.; Hernández-Galán, R.; Collado, I.G. Global antifungal profile optimization of chlorophenyl derivatives against *Botrytis cinerea* and *Colletotrichum gloeosporioides*. *J. Agric. Food Chem.* **2009**, *57*, 4838–4843. [CrossRef] [PubMed]
37. Gao, S.; Xu, Z.; Wang, X.; Feng, H.; Wang, L.; Zhao, Y.; Wang, Y.; Tang, X. Synthesis and antifungal activity of aspirin derivatives. *Asian J. Chem.* **2014**, *26*, 7157–7159. [CrossRef]
38. Chen, C.-J.; Song, B.-A.; Yang, S.; Xu, G.-F.; Bhadury, P.S.; Jin, L.-H.; Hu, D.-Y.; Li, Q.-Z.; Liu, F.; Xue, W.; et al. Synthesis and antifungal activities of 5-(3,4,5-trimethoxyphenyl)-2-sulfonyl-1,3,4-thiadiazole and 5-(3,4,5-trimethoxyphenyl)-2-sulfonyl-1,3,4-oxadiazole derivatives. *Bioorg. Med. Chem.* **2007**, *15*, 3981–3989. [CrossRef] [PubMed]
39. Herth, W.; Schnepf, E. The fluorochrome, calcofluor white, binds oriented to structural polysaccharide fibrils. *Protoplasma* **1980**, *105*, 129–133. [CrossRef]
40. Bolton, J.L.; Trush, M.A.; Penning, T.M.; Dryhurst, G.; Monks, T.J. Role of quinones in toxicology. *Chem. Res. Toxicol.* **2000**, *13*, 135–160. [CrossRef] [PubMed]
41. Dahlin, D.C.; Miwa, G.T.; Lu, A.Y.H.; Nelson, S.D. *N*-acetyl-*p*-benzoquinone imine: A cytochrome P-450-mediated oxidation product of acetaminophen. *Proc. Natl. Acad. Sci. USA* **1984**, *81*, 1327–1331. [CrossRef] [PubMed]
42. Mitchell, J.R.; Jollow, D.J.; Potter, W.Z.; Davis, D.C.; Gillette, J.R.; Brodie, B.B. Acetaminophen-induced hepatic necrosis. *J. Pharmacol. Exp. Ther.* **1973**, *187*, 185–194. [PubMed]
43. Pócsi, I.; Prade, R.A.; Penninckx, M.J. Glutathione, altruistic metabolite in fungi. *Adv. Microb. Physiol.* **2004**, *49*, 1–76. [PubMed]
44. Lopez, S.N.; Castelli, M.V.; Zacchino, S.A.; Domínguez, J.N.; Lobo, G.; Charris-Charris, J.; Cortes, J.C.; Ribas, J.C.; Devia, C.; Rodriguez, A.M.; et al. In vitro antifungal evaluation and structure–activity relationships of a new series of chalcone derivatives and synthetic analogues, with inhibitory properties against polymers of the fungal cell wall. *Bioorg. Med. Chem.* **2001**, *9*, 1999–2013. [CrossRef]
45. Muñoz, G.; Hinrichsen, P.; Brygoo, Y.; Giraud, T. Genetic characterisation of *Botrytis cinerea* populations in Chile. *Mycol. Res.* **2002**, *106*, 594–601. [CrossRef]

Sample Availability: Samples of the compounds **3a–j** are available from the authors.

© 2019 by the authors. Licensee MDPI, Basel, Switzerland. This article is an open access article distributed under the terms and conditions of the Creative Commons Attribution (CC BY) license (http://creativecommons.org/licenses/by/4.0/).

Article

A Convenient, Rapid, Sensitive, and Reliable Spectrophotometric Assay for Adenylate Kinase Activity

Kai Song [1], Yejing Wang [2,*], Yu Li [1], Chaoxiang Ding [1], Rui Cai [2], Gang Tao [1], Ping Zhao [1], Qingyou Xia [1] and Huawei He [1,3,*]

[1] Biological Science Research Center, Southwest University, Beibei, Chongqing 400715, China; Kaisong@email.swu.edu.cn (K.S.); liyu315@swu.edu.cn (Y.L.); ding7197767@email.swu.edu.cn (C.D.); taogang@email.swu.edu.cn (G.T.); zhaop@swu.edu.cn (P.Z.); xiaqy@swu.edu.cn (Q.X.)
[2] State Key Laboratory of Silkworm Genome Biology, College of Biotechnology, Southwest University, Beibei, Chongqing 400715, China; cairui0330@email.swu.edu.cn
[3] Chongqing Key Laboratory of Sericultural Science, Chongqing Engineering and Technology Research Center for Novel Silk Materials, Southwest University, Beibei, Chongqing 400715, China
* Correspondence: yjwang@swu.edu.cn (Y.W.); hehuawei@swu.edu.cn (H.H.); Tel.: +86-23-6825-1575 (Y.W. & H.H.)

Academic Editor: Stefano Serra
Received: 19 January 2019; Accepted: 12 February 2019; Published: 13 February 2019

Abstract: Enzymatic activity assays are essential and critical for the study of enzyme kinetics. Adenylate kinase (Adk) plays a fundamental role in cellular energy and nucleotide homeostasis. To date, assays based on different principles have been used for the determination of Adk activity. Here, we show a spectrophotometric analysis technique to determine Adk activity with bromothymol blue as a pH indicator. We analyzed the effects of substrates and the pH indicator on the assay using orthogonal design and then established the most optimal assay for Adk activity. Subsequently, we evaluated the thermostability of Adk and the inhibitory effect of KCl on Adk activity with this assay. Our results show that this assay is simple, rapid, and precise. It shows great potential as an alternative to the conventional Adk activity assay. Our results also suggest that orthogonal design is an effective approach, which is very suitable for the optimization of complex enzyme reaction conditions.

Keywords: enzymatic activity assay; adenylate kinase; spectrophotometry; orthogonal experiment; bromothymol blue

1. Introduction

Adenylate kinase (Adk; ATP:AMP phosphotransferase, EC 2.7.4.3), also known as myokinase, is a conserved phosphoryl transferase, which catalyzes the translocation of a phosphoryl group between nucleotides in the reversible reaction (AMP + Mg^{2+}•ATP  Mg^{2+}•ADP + ADP) [1]. Adk is ubiquitous in different tissues of all living systems and plays a fundamental role in cellular energy and nucleotide homeostasis. The hydrogen bond between the adenine moiety and the backbone of Adk is critical for ATP selectivity and can help Adk recognize the correct substrates in the complex cellular environment [2]. Adk is involved in the regulation of cell differentiation, maturation, apoptosis, and oncogenesis. Adk mutations in humans cause a severe disease called reticular dysgenesis [3]. Adk is regarded as a potential target for medical diagnosis and treatment due to its close correlation with other diseases, such as aleukocytosis, hemolytic anemia, and primary ciliary dyskinesia [4]. To date, three methods have been proposed to determine Adk activity based on the detectable changes accompanied with this reaction, such as light absorption, acidity, or the coupled reaction products [5]. The manometric assay, established by Colowick and Kalckar [6], is used for the detection of Adk by measuring CO_2 liberation from a bicarbonate buffer. The reaction catalyzed by Adk is coupled to hexokinase, which specifically catalyzes the transformation of the terminal phosphate from ADP to glucose. The overall reaction is as follows:

$$ATP + AMP \xrightleftharpoons{Adk} ADP + ADP \quad (1)$$

$$ADP + glucose \xrightarrow{hexokinase} glucose-6-P + AMP + H^+ \quad (2)$$

In the presence of excess hexokinase, the reaction rate is proportional to the Adk concentration. The forward direction reaction is defined as the formation of AMP and ATP from two ADPs, and the reverse direction is defined as the formation of two ADPs from AMP and ATP. Adk activity is conventionally measured in vitro by a spectrophotometric assay. For the forward direction reaction, Chiu et al. [7] have developed a modified assay of Oliver [8] to determine Adk activity by coupling the reaction to hexokinase and glucose-6-phosphate dehydrogenase in which the final product, NADPH, is measured spectrophotometrically at 340 nm. The overall reaction is as follows: H^+

$$ADP + ADP \xrightleftharpoons{Adk} ATP + AMP \quad (3)$$

$$ATP + glucose \xrightarrow{hexokinase} glucose-6-P + ADP \quad (4)$$

$$glucose-6-P + NADP \xrightarrow{glucose-6-phosphate\ dehydrogenase} 6-phosphogluconic\ acid + NADPH \quad (5)$$

Adk activity is also measured in the reverse direction by coupling the reaction to pyruvate kinase and lactate dehydrogenase and measuring the oxidation of NADH at 340 nm [5]. The principle of the assay is as follows:

$$ATP + AMP \xrightleftharpoons{Adk} ADP + ADP \quad (6)$$

$$Phosphoenolpyruvate + ADP \xrightarrow{Pyruvate\ kinase} ATP + Pyruvate \quad (7)$$

$$Pyruvate + NADH + H + \xrightarrow{Lactic\ dehydrogenase} Lactate + NAD^+ \quad (8)$$

These assays have been used to determine Adk activity for the past decades. However, some disadvantages are also obvious for these assays. Firstly, these assays are time-consuming, multistep processes that require the assistance of other enzymes and are easily subject to errors at each step. Secondly, it is difficult to study the effects of activators and inhibitors on Adk activity with the aid of other enzymes. Finally, the real initial rate of Adk reaction cannot be determined accurately [9]. Therefore, it is necessary to develop a more convenient and accurate assay for Adk activity in vitro.

Acid–base indicators are usually applied in enzymatic assay for their extraordinary sensitivity to pH change. In 2002, Yu et al. [10] established an arginine kinase activity assay based on the light absorption of a complex acid–base indicator consisting of thymol blue and cresol red. In the reaction catalyzed by arginine kinase, the produced protons resulted in a decrease in pH of the reaction mixture, thus reducing the absorbance of the mixed indicator in the solution at 575 nm. The arginine kinase activity could be determined according to the change of the absorbance at 575 nm. In the same way, Dhale et al. developed a rapid and sensitive assay to measure L-asparaginase activity with methyl red as an indicator [11]. Bromothymol blue is an excellent indicator as it forms a highly conjugated structure while deprotonated in alkaline solution, resulting in an obvious color change from yellow to blue and the corresponding absorbance change [12].

Enzyme activity is typically influenced by many factors. The traditional method is to do a multifactor analysis that tests all possible combinations of the different factors. However, this takes up a lot of time and resources as the number of full factorial experiments is very large. As an alternative, the orthogonal design method has been proposed and established. The orthogonal experimental design [13] is a multifactor experiment design assay. It selects representative samples from a full factorial assay in a way that the samples are distributed uniformly within the test range, thus representing the overall situation. Therefore, it is highly efficient for the arrangement of multifactor experiments with optimal combination levels. The orthogonal design has three advantages: (1) The number of tests required to complete the experiment is relatively small. (2) The data points are evenly distributed. (3) The test results can be analyzed by mathematical calculation (e.g., range analysis and variance analysis), which is particularly useful to quantify the results.

In this study, we developed a one-step assay for Adk activity. It is based on proton generation after the addition of ATP and AMP as the substrates, which can be measured spectrophotometrically at 614 nm using bromothymol blue as a pH indicator. We investigated four factors affecting Adk activity—ATP, AMP, bromothymol blue, and glycine–NaOH buffer—at three levels and determined the best combination for Adk activity assay by an orthogonal experimental design. Finally, we evaluated the thermostability of Adk and the inhibitory effect of KCl on Adk activity with this assay. Our results suggest that this assay is simple, precise, less expensive, and a potential alternative to the conventional enzymes-coupled assay extensively used in clinical and research laboratories.

2. Results

In this study, Adk activity was determined by a direct and continuous spectrophotometric technique without coupled enzymes. In the enzymatic reaction catalyzed by Adk, the formation of two ADPs from ATP and AMP is accompanied by the generation of hydrogen ions. Bromothymol blue is an excellent acid–base indicator as it forms a highly conjugated structure while protonated in acid solution, resulting in an obvious color change from blue to yellow. The absorbance of bromothymol blue at 614 nm is associated with the hydrogen ion concentration in solution. Thus, Adk activity can be monitored in real time by the absorbance of bromothymol blue at 614 nm in solution, which can be detected by a sensitive spectrophotometer. The principle of this assay is illustrated in Figure 1.

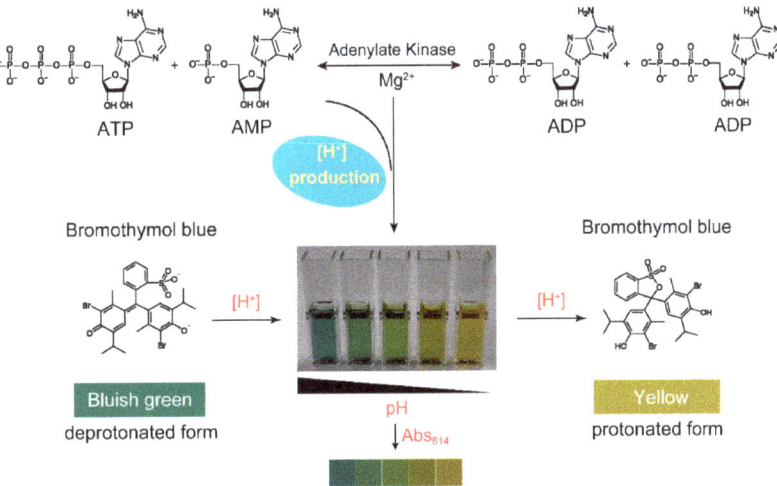

Figure 1. The principle of the spectrophotometric assay for adenylate kinase (Adk) activity.

The effects of substrates and the pH indicator on the assay were analyzed using an orthogonal design, which is key to establishing the most optimal assay for Adk activity. The factors and levels affecting Adk activity assay are shown in Table 1.

Table 1. Factors and levels affecting Adk activity assay.

Level	A ATP (mM)	B AMP (mM)	C Bromothymol Blue (mM)	D Glycine–NaOH (mM)
1	2.0	1.0	0.0930	0.1
2	2.5	1.5	0.1084	0.3
3	3.0	2.0	0.1238	0.5

2.1. The Maximum Absorption Wavelength of Reaction Mixture

A set of nine tests designed by orthogonal experiment is shown in Table 2. The absorption spectrum of each set was scanned from 450 to 800 nm in the presence of 5 mM MgAC$_2$. All spectra showed the same maximum absorption located at 614 nm without shift, as shown in Figure 2a.

Table 2. Orthogonal array (9, 3^4) for the analysis of the effects of ATP, AMP, bromothymol blue, and glycine–NaOH buffer on the Adk activity assay.

No.	Combination	Factor				ΔAbs [a] (0–30 s)
		A	B	C	D	
1	$A_1B_1C_1D_1$	2.0	1.0	0.0930	0.1	0.0914 ± 0.0020
2	$A_1B_2C_2D_2$	2.0	1.5	0.1084	0.3	0.0432 ± 0.0045
3	$A_1B_3C_3D_3$	2.0	2.0	0.1238	0.5	0.0112 ± 0.0017
4	$A_2B_1C_2D_3$	2.5	1.0	0.1084	0.5	0.0404 ± 0.0017
5	$A_2B_2C_3D_1$	2.5	1.5	0.1238	0.1	0.0221 ± 0.0020
6	$A_2B_3C_1D_2$	2.5	2.0	0.0930	0.3	0.0606 ± 0.0022
7	$A_3B_1C_3D_2$	3.0	1.0	0.1238	0.3	0.0341 ± 0.0009
8	$A_3B_2C_1D_3$	3.0	1.5	0.0930	0.5	0.0705 ± 0.0049
9	$A_3B_3C_2D_1$	3.0	2.0	0.1084	0.1	0.0382 ± 0.0017
T_1		0.1458	0.1659	0.2225	0.1517	
T_2		0.1231	0.1358	0.1218	0.1379	
T_3		0.1428	0.1100	0.0674	0.1221	
t_1		0.0486	0.0553	0.0742	0.0506	
t_2		0.0410	0.0453	0.0406	0.0460	
t_3		0.0476	0.0367	0.0225	0.0407	
	Range (R)	0.0076	0.0186	0.0517	0.0099	
	Order		C > B > D > A			
	Optimal level	A_1	B_1	C_1	D_1	
	Optimal combination		$A_1B_1C_1D_1$			

[a] Arithmetic mean of the absorbance changes of bromothymol blue at 614 nm (0–30 s) of three independent tests at each level under the same factor. Ti (T1, T2, T3) is the sum of the recorded absorbance changes (ΔAbs) at the same level and under the same factor, and ti (t1, t2, t3) is the arithmetic mean of Ti. The mean of ti represents the influence of different levels under the same factor on the absorbance of bromothymol blue. Range (R) is the difference between the maximum and the minimum of ti, indicating the effect of each factor on the absorbance of bromothymol blue. The greater the R value, the greater is the influence of this factor on the absorbance of bromothymol blue at 614 nm.

2.2. Optimization of Adk Activity Assay

The effects of ATP, AMP, bromothymol blue, and glycine–NaOH buffer on the Adk activity were analyzed, as shown in Table 2. The significance of these factors on the assay was determined by the range (R) value listed in Table 2. The results showed that the order of R value was RC > RB > RD > RA. Hence, the significance order of these factors on the assay was C > B > D > A, namely, bromothymol blue > AMP > glycine–NaOH buffer > ATP. Similarly, the significance of different levels on the assay was determined by the ti value listed in Table 2. The results showed that the orders were t1 > t3 > t2 for factor A, t1 > t2 > t3 for factor B, t1 > t2 > t3 for factor C, and t1 > t2 > t3 for factor D. Thus, the most optimal combination for Adk activity assay was A1B1C1D1, which was composed of 2 mM ATP, 1 mM AMP, 0.093 mM bromothymol blue, and 0.1 mM glycine–NaOH buffer.

Next, we conducted a full factorial design for two primary factors of bromothymol blue and AMP at three levels to further optimize the assay. The design matrix and parameters are listed in Table 3. The results showed that AB3C1D had the most significant absorbance change at 614 nm among all the tested conditions (Table 4), indicating AB3C1D (2 mM ATP, 0.6 mM AMP, 0.093 mM bromothymol blue, and 0.1 mM glycine-NaOH buffer) was the most optimal reaction condition for the Adk activity assay.

Table 3. Factors and levels affecting Adk activity assay with the constants of A and D.

Level	A ATP (mM)	B AMP (mM)	C Bomothymol Blue (mM)	D Glycine–NaOH (mM)
1	2.0	1.0	0.0930	0.1
2	2.0	0.8	0.0775	0.1
3	2.0	0.6	0.0620	0.1

Table 4. A full factorial design for the analysis of the effects of AMP, bromothymol blue on the Adk activity assay with the constants of A and D.

Run	Combination	Factor				ΔAbs [b] (0–30 s)
		A	B	C	D	
1	AB_1C_1D	2.0	1.0	0.0930	0.1	0.0914 ± 0.0020
2	AB_1C_2D	2.0	1.0	0.0775	0.1	0.0926 ± 0.0030
3	AB_1C_3D	2.0	1.0	0.0620	0.1	0.0764 ± 0.0030
4	AB_2C_1D	2.0	0.8	0.0930	0.1	0.0914 ± 0.0020
5	AB_2C_2D	2.0	0.8	0.0775	0.1	0.0930 ± 0.0018
6	AB_2C_3D	2.0	0.8	0.0620	0.1	0.0680 ± 0.0041
7	AB_3C_1D	2.0	0.6	0.0930	0.1	0.0983 ± 0.0028
8	AB_3C_2D	2.0	0.6	0.0775	0.1	0.0850 ± 0.0054
9	AB_3C_3D	2.0	0.6	0.0620	0.1	0.0604 ± 0.0001

[b] Arithmetic mean of the absorbance changes of bromothymol blue at 614 nm (0–30 s) of three independent tests at each level under the same factor.

2.3. Effect of H^+ on the Absorbance of Bromothymol Blue

To determine the sensitivity of bromothymol blue on the Adk activity assay, we measured the response of bromothymol blue to hydrogen ion under the most optimal condition, AB3C1D. The absorbance spectrum of bromothymol blue was scanned from 450 to 800 nm in the presence of various concentrations of HCl. The results showed that the absorbance of bromothymol blue at 614 nm gradually declined with the increase in hydrogen ion concentration (Figure 2b). The absorbance change can be linearly fitted as the function of hydrogen ion concentration with the following equation (Figure 2c):

$$y = 0.138 * x - 0.066$$

The results suggested that the absorbance of bromothymol blue at 614 nm had a good response to pH change in the assay, and the absorbance change was positively correlated with the hydrogen ion concentration.

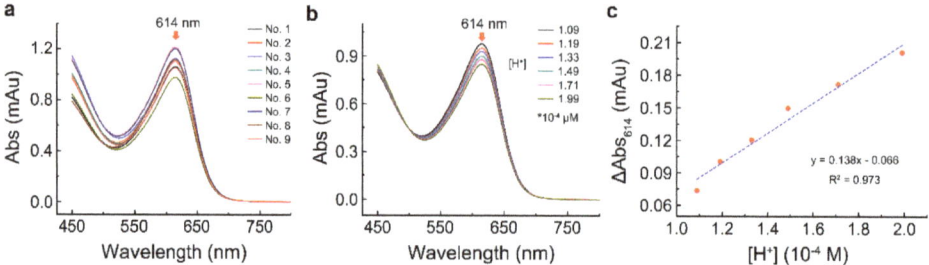

Figure 2. The maximum absorption wavelength of the reaction mixture and its relationship with the hydrogen ion concentration. (**a**) The absorption spectra of nine different combinations in the presence of 5 mM MgAC$_2$; (**b**) effect of hydrogen ion concentration on the absorption of the reaction system; (**c**) the correlation of the absorbance change of bromothymol blue at 614 nm with hydrogen ion concentration.

2.4. Effect of Adk Contents on the Reaction Velocity

The effect of Adk contents on the reaction velocity was determined as described in the Materials and Methods section. The results showed that, with the increase in Adk contents, the reaction velocity increased, and the reaction time required to reach equilibrium shortened (Figure 3a). In the first 5 s, the absorbance of bromothymol blue at 614 nm (Abs$_{614}$) declined linearly with time; thus, the slope of the reaction in the first 5 s was defined as the initial reaction velocity. The plot in Figure 3b shows that the absorbance change of bromothymol blue at 614 nm could be linearly fitted as the function of Adk contents.

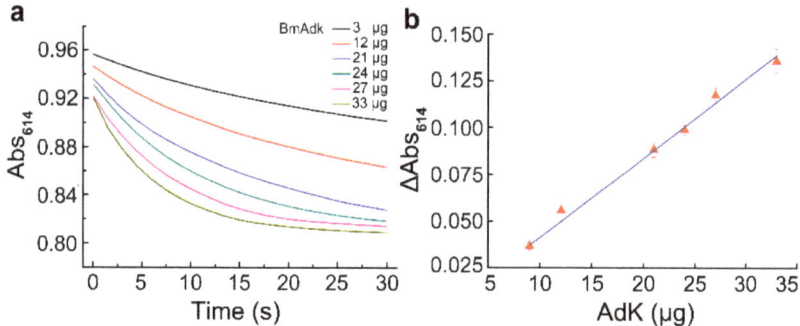

Figure 3. The effect of Adk contents on the assay. (**a**) Effect of different Adk contents on the assay; (**b**) the correlation of the absorbance change of bromothymol blue at 614 nm with different Adk contents.

2.5. Effect of Temperature and KCl on Adk Activity

The effect of temperature on Adk activity was investigated to characterize the thermostability of Adk. The results showed that Adk activity was almost unaffected under 45 °C. However, when the temperature was increased to 60 °C, Adk quickly lost its activity (Figure 4a). Adk from the muscle has a half-life of 30 min in 0.1 N hydrochloric acid at 100 °C [14]. Our results indicate that the thermostability of Adk from *Bombyx mori* (BmAdk) is much lower than that of Adk from the muscle.

Allan Hough et al. proved that KCl can almost completely inhibit myokinase activity [15]. Here, the effect of KCl on Adk activity was assessed with the assay. The results showed that low concentration of KCl (<5 mM) had a slight inhibitory effect on Adk activity. With the increase in KCl concentration, the inhibitory effect of KCl on Adk activity became more and more obvious. About 70 mM KCl resulted in 50% loss of Adk activity (Figure 4b). Compared with the "three-minute" method [15], the inhibitory effect of KCl on Adk activity could be assessed more easily with our developed assay.

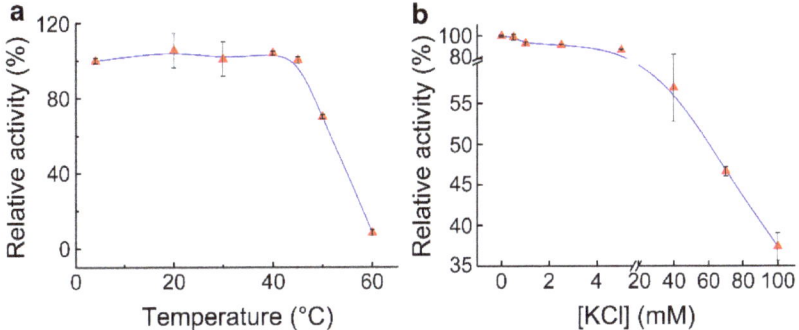

Figure 4. The thermostability of Adk and the inhibition of KCl on Adk activity. (**a**) Effect of temperature on Adk activity; (**b**) effect of KCl concentration on Adk activity.

3. Materials and Methods

3.1. Chemicals and Materials

ATP and AMP were purchased from Aladdin (Shanghai, China) in the form of sodium salt. Magnesium acetate was from Sigma (St. Louis, MO, USA). Bromothymol blue sodium salt, glycine, and other reagents all came from Sangon Biotech Corp. (Shanghai, China). Plastic cuvettes were purchased from Centome Corp. (Chengdu, China).

3.2. Adk Preparation and Concentration Determination

The DNA fragment encoding Adk was obtained from the cDNA library by PCR from the midgut of *Bombryx mori* strain Dazao using primer sets (5'-ATGGCACCGGCCGCTGC-3' and 5'-TTACAAA GCAGACCGTGCTCTGCTG-3'). The amplification product was gel-purified, recovered, and inserted into plasmid vectors pSKB2. The bacterial transformants containing error-free inserts were identified. Adk was expressed in *Escherichia coli* BL21(DE3) and purified by Ni-NTA affinity chromatography (GE Healthcare, Chicago, IL, USA). The fused polyhistidine tag was cleaved by Prescission protease (GE Healthcare, USA) and removed as described by Liu et al. [16]. Protein concentration was determined using the extinction coefficient of 12,950 $M^{-1} \cdot L \cdot cm^{-1}$ at 280 nm on a NanoDrop 2000C spectrophotometer (Thermo Fisher, Waltham, MA, USA).

3.3. Adk Activity Assay

ATP and AMP were used as the substrates in the assay. The reaction mixture (1 mL) was composed of 2 mM ATP, 0.6 mM AMP, 0.1 mM glycine–NaOH (pH 9.0), 0.093 mM bromothymol blue, and 5 mM $MgAC_2$. The initial absorbance of the freshly prepared reaction mixture at 614 nm was adjusted to 1.05 with approximately 0.5 M NaOH so that the absorbance of the mixture would not be changed after the addition of Adk. The purified Adk was exchanged into buffer A (20 mM Tris-HCl, pH 7.6, 150 mM NaCl, 5% glycerol, and 0.1 mM dithiothreitol (DTT)) via gel filtration. The reaction was triggered by adding 15 µg Adk into the mixture. The reaction velocity was defined as the slope of the absorbance change of bromothymol blue at 614 nm in the initial 30 s, which was recorded on a DU 800 nucleic acid/protein analyzer (BeckmanCoulter, Brea, CA, USA) using a 1-cm light path plastic cuvette. For the control reaction, Adk was replaced with buffer A. All measurements were carried out at 25 °C. Each test was replicated at least three times.

3.4. Orthogonal Design

The concentration of ATP, AMP, bromothymol blue, and glycine–NaOH buffer is vital for Adk activity assay. Consequently, these factors were considered in the orthogonal design to screen the most optimal conditions for Adk activity. Assuming there were three levels for each factor, and the interaction among the factors were not taken into account, the orthogonal experiment of four factors at three levels was designed.

4. Discussion

Adk plays a crucial role in maintaining a balance of cellular energy and nucleic acid metabolism. Human Adk isoenzymes specifically expressed in organs are regarded as important indicators of organ dysfunction [17] or differentiation stages [18]. The Adk activity assays that have been developed so far are largely dependent on the coupled secondary enzymes hexokinase and glucose-6-phosphate dehydrogenase or adenosine monophosphate deaminase [19]. In addition, the different conditions between the coupled enzymes and Adk affect the application of these assays [14]. Therefore, developing a convenient, rapid, sensitive, reliable, and economic assay for Adk activity is of great significance.

Here, we developed a spectrophotometric assay for Adk activity using bromothymol blue as a pH indicator without coupled enzymes. The effective range of bromothymol blue is pH 6.0–7.6. Adk is optimally active at pH 7.6 [14]. Hence, Adk was dissolved in pH 7.6 buffer to keep it active. The correspondence of pH and the absorbance of the system are listed in Table 5. Here, we set the initial absorbance of 1.05 as the beginning of the reaction, which represented pH 6.8 of the reaction system. When 5 μL buffer was added into the system (995 μL), there was almost no change in the pH of the system. At the same time, the assay system could ensure bromothymol blue had a sensitive and stable response to pH change caused by Adk-catalyzed reaction.

Table 5. Correspondence between pH and the absorbance of bromothymol blue at 614 nm.

pH	6.47	6.54	6.73	6.79	6.85	7.07	7.21	7.28
Abs	0.8145	0.9197	0.9438	1.0375	1.0847	1.2385	1.3850	1.4188

The effects of ATP, AMP, bromothymol blue, and glycine–NaOH buffer on the Adk activity assay were determined by a rational orthogonal design. For a full factorial assay with four factors at three levels, the number of tests is up to 81 (3^4). However, our rational orthogonal design greatly reduced the unnecessary experiments and effectively achieved significant results (Table 2). The orthogonal experimental design is a rational design method for multifactor experiment, which selects representative points from a full factorial assay to represent the overall situation. Therefore, it is highly efficient for the design of multifactor experiments with optimal combination levels. By means of the orthogonal design, we greatly reduced the number of required experiments and achieved significant results with the least number of experiments.

The optimal temperature for the growth of *Bombyx mori* is about 25 °C. Adk from *Bombyx mori* was relatively stable below 45 °C, implying its significance on the growth and development of *Bombyx mori*. The Adk structure suggests that the hydrophobic core packing is important for the stability and activity of Adk [20]. AMP has a strong inhibition effect on the reaction with ADP as the substrate, thus resulting in a rapid decrease in the reaction rate with time [6,19]. Here, we found that high concentration of AMP could inhibit Adk activity, which is in line with a previous report. Slater et al. [21] found that the inhibitory effect of AMP was easily observed even when excessive hexokinase and glucose were present. The inhibition was ascribed to the affinity of ADP to Adk, which is lower than that of AMP with Adk. However, in this study, the substrates were ATP and AMP. Therefore, the inhibition of AMP on Adk activity may be attributed to the noncompetitive inhibition of AMP with respect to ATP [22].

To summarize, we developed a simple and rapid assay to determine Adk activity with bromothymol blue as an indicator instead of coupled enzymes. The assays that have been developed so far require the assistance of other enzymes to convert the reaction product to other detectable signals; thus, they consist of multiple reaction steps and are discontinuous. The assays are not accurate as they are easily subject to errors at each step. However, this new assay only relies on the protons produced in the reaction and the corresponding absorbance changes of bromothymol blue in the reaction solution. Small changes in pH can lead to significant changes in the absorbance of bromothymol blue at 614 nm. It does not need the assistance of any other enzyme. Bromothymol blue is less expensive than the coupled enzymes, and the assay can be done in just one step, thus reducing the chance of errors and improving reliability. Compared with the existing assays, it is more simple, sensitive, and precise. In addition, it can be applied to research the activation or inhibition of Adk as it is continuous. However, although this assay is simple and precise, the reaction substrate must be prepared freshly as carbon dioxide in the air may interfere with the assay. ATP spontaneously and slowly hydrolyzes in solution, which can also cause an interference on the assay. Nevertheless, it can be a good alternative to the conventional enzymes-coupled Adk activity assay extensively used in clinical and research laboratories.

Author Contributions: K.S., Y.W., Y.L., C.D., and H.H. conceived and designed the experiments; K.S. and Y.L. performed the experiments; K.S., Y.L., and Y.W. analyzed the data; R.C., T.G., and P.Z. contributed reagents/materials/analysis tools; K.S. and Y.W. wrote the draft; Y.W. and H.H. supervised the research; Q.X. and H.H. revised the manuscript.

Funding: This work was supported by the National Natural Science Foundation of China (31572465), the State Key Program of the National Natural Science of China (31530071), Fundamental Research Funds for the Central Universities (XDJK2018B010, XDJK2018C063), start-up grant from Southwest University (SWU112111), Graduate Research and Innovation Project of Chongqing (CYB17069), and the Open Project Program of Chongqing Engineering and Technology Research Center for Novel Silk Materials (silkgczx2016003).

Conflicts of Interest: The authors declare no conflict of interest. The founding sponsors had no role in the design of the study; in the collection, analyses, or interpretation of data; in the writing of the manuscript; or in the decision to publish the results.

Abbreviations

Adk	Adenylate kinase
ATP	adenosine 5′-triphosphates
ADP	adenosine 5′-diphosphates
AMP	adenosine 5′-monophosphate
NADPH	reduced nicotinamide adenine dinucleotide phosphate
Abs	absorbance

References

1. Krishnamurthy, H.; Lou, H.F.; Kimple, A.; Vieille, C.; Cukier, R.I. Associative mechanism for phosphoryl transfer: A molecular dynamics simulation of Escherichia coli adenylate kinase complexed with its substrates. *Proteins* **2005**, *58*, 88–100. [CrossRef] [PubMed]
2. Rogne, P.; Rosselin, M.; Grundstrom, C.; Hedberg, C.; Sauer, U.H.; Wolf-Watz, M. Molecular mechanism of ATP versus GTP selectivity of adenylate kinase. *Proc. Natl. Acad. Sci. USA* **2018**, *115*, 3012–3017. [CrossRef]
3. Oshima, K.; Saiki, N.; Tanaka, M.; Imamura, H.; Niwa, A.; Tanimura, A.; Nagahashi, A.; Hirayama, A.; Okita, K.; Hotta, A.; et al. Human AK2 links intracellular bioenergetic redistribution to the fate of hematopoietic progenitors. *Biochem. Biophys. Res. Commun.* **2018**, *497*, 719–725. [CrossRef]
4. Wujak, M.; Czarnecka, J.; Gorczycka, M.; Hetmann, A. Human adenylate kinases-classification, structure, physiological and pathological importance. *Postep. Hig. Med. Dosw.* **2015**, *69*, 933–945. [CrossRef] [PubMed]
5. Bucher, T.; Pfleiderer, G. Pyruvate Kinase from Muscle. *Method Enzymol.* **1955**, *1*, 435–440.
6. Colowick, S.P.; Kalckar, H.M. The role of myokinase in transphosphorylations I. The enzymatic phosphorylation of hexoses by adenyl pyrophosphate. *J. Biol. Chem.* **1943**, *148*, 117–126.
7. Chiu, C.S.; Su, S.; Russell, P.J. Adenylate kinase from baker's yeast. I. Purification and intracellular location. *Biochimica Biophys. Acta* **1967**, *132*, 361–369. [CrossRef]
8. Oliver, I.T. Spectrophotometric Method for the Determination of Creatine Phosphokinase and Myokinase. *Biochem. J.* **1955**, *61*, 116–122. [CrossRef] [PubMed]
9. Haslam, R.J., Mills, D.C.B. Adenylate Kinase of Human Plasma Erythrocytes and Platelets in Relation to Degradation of Adenosine Diphosphate in Plasma. *Biochem. J.* **1967**, *103*, 773. [CrossRef] [PubMed]

10. Yu, Z.; Pan, J.; Zhou, H.M. A direct continuous PH-spectrophotometric assay for arginine kinase activity. *Protein Peptide Lett.* **2002**, *9*, 545–552.
11. Dhale, M.A.; Mohan-Kumari, H.P. A comparative rapid and sensitive method to screen L-asparaginase producing fungi. *J. Microbiol. Meth.* **2014**, *102*, 66–68. [CrossRef] [PubMed]
12. De Meyer, T.; Hemelsoet, K.; Van Speybroeck, V.; De Clerck, K. Substituent effects on absorption spectra of pH indicators: An experimental and computational study of sulfonphthaleine dyes. *Dyes Pigments* **2014**, *102*, 241–250. [CrossRef]
13. Zhu, J.J.; Chew, D.A.S.; Lv, S.N.; Wu, W.W. Optimization method for building envelope design to minimize carbon emissions of building operational energy consumption using orthogonal experimental design (OED). *Habitat. Int.* **2013**, *37*, 148–154. [CrossRef]
14. Colowick, S.P. Adenylate Kinase (Myokinase, Adp Phosphomutase). *Method Enzymol.* **1955**, *2*, 598–604.
15. Bowen, W.J.; Kerwin, T.D. The Kinetics of Myokinase. 1. Studies of the Effects of Salts and Ph and of the State of Equilibrium. *Arch. Biochem. Biophys.* **1954**, *49*, 149–159. [CrossRef]
16. Liu, L.; Li, Y.; Wang, Y.; Zhao, P.; Wei, S.; Li, Z.; Chang, H.; He, H. Biochemical characterization and functional analysis of the POU transcription factor POU-M2 of Bombyx mori. *Int. J. Biol. Macromol.* **2016**, *86*, 701–708. [CrossRef] [PubMed]
17. Bernstei, L.H.; Horenstein, J.M.; Sybers, H.D.; Russell, P.J. Adenylate Kinase in Human Tissue. 2. Serum Adenylate Kinase and Myocardial-Infarction. *J. Mol. Cell Cardiol.* **1973**, *5*, 71–85. [CrossRef]
18. Russell, P.J.; Horenstein, J.M.; Goins, L.; Jones, D.; Laver, M. Adenylate Kinase in Human Tissues. 1. Organ Specificity of Adenylate Kinase Isoenzymes. *J. Biol. Chem.* **1974**, *249*, 1874–1879.
19. Kalckar, H.M. The role of myokinase in transphosphorylations II. The enzymatic action of myokinase on adenine nucleotides. *J. Biol. Chem.* **1943**, *148*, 127–137.
20. Moon, S.; Kim, J.; Bae, E. Structural analyses of adenylate kinases from Antarctic and tropical fishes for understanding cold adaptation of enzymes. *Sci. Rep.* **2017**, *7*, 16027. [CrossRef]
21. Slater, E.C. A Method of Measuring the Yield of Oxidative Phosphorylation. *Biochem. J.* **1953**, *53*, 521–530. [CrossRef]
22. Palella, T.D.; Andres, C.M.; Fox, I.H. Human Placental Adenosine Kinase-Kinetic Mechanism and Inhibition. *J. Biol. Chem.* **1980**, *255*, 5264–5269. [PubMed]

Sample Availability: Not available.

© 2019 by the authors. Licensee MDPI, Basel, Switzerland. This article is an open access article distributed under the terms and conditions of the Creative Commons Attribution (CC BY) license (http://creativecommons.org/licenses/by/4.0/).

Article

Fungi-Mediated Biotransformation of the Isomeric Forms of the Apocarotenoids Ionone, Damascone and Theaspirane

Stefano Serra * and Davide De Simeis

C.N.R. Istituto di Chimica del Riconoscimento Molecolare, Via Mancinelli 7, 20131 Milano, Italy; dav.biotec01@gmail.com
* Correspondence: stefano.serra@cnr.it or stefano.serra@polimi.it; Tel.: +39-02-2399-3076

Received: 19 November 2018; Accepted: 20 December 2018; Published: 21 December 2018

Abstract: In this work, we describe a study on the biotransformation of seven natural occurring apocarotenoids by means of eleven selected fungal species. The substrates, namely ionone (α-, β- and γ-isomers), 3,4-dehydroionone, damascone (α- and β-isomers) and theaspirane are relevant flavour and fragrances components. We found that most of the investigated biotransformation reactions afforded oxidized products such as hydroxy- keto- or epoxy-derivatives. On the contrary, the reduction of the keto groups or the reduction of the double bond functional groups were observed only for few substrates, where the reduced products are however formed in minor amount. When starting apocarotenoids are isomers of the same chemical compound (e.g., ionone isomers) their biotransformation can give products very different from each other, depending both on the starting substrate and on the fungal species used. Since the majority of the starting apocarotenoids are often available in natural form and the described products are natural compounds, identified in flavours or fragrances, our biotransformation procedures can be regarded as prospective processes for the preparation of high value olfactory active compounds.

Keywords: biotransformation; oxidation; apocarotenoids; flavours; fungi; ionone; damascone; theaspirane

1. Introduction

In Nature, the oxidative degradation of the conjugated tetraterpene carotenoids (C_{40}) produces a plethora of smaller derivatives, called apocarotenoids [1], which possess a range of different chemical structures and biological activities. Among these natural products, compounds having thirteen carbon atoms in their frameworks are relevant flavours or fragrances and their manufacturing represents an important economic resource for chemical companies [2]. The combination of the great diversity of the carotenoids chemical structures with the different possible degradation pathways, gives rise to a huge number of flavours and fragrances.

It is worth noting that the primary odorous C_{13} apocarotenoids, namely ionone, damascone and theaspirane isomers (Figure 1), are cyclohexene derivatives and the possibility of three different positions of the double bond, the presence of a stereogenic center in position 6 (carotenoids numbering) and the eventual structural rearrangements, can give rise to a large number of isomers. In addition, the latter volatile compounds are often accompanied by structural related apocarotenoids having further oxygen atoms in their chemical framework.

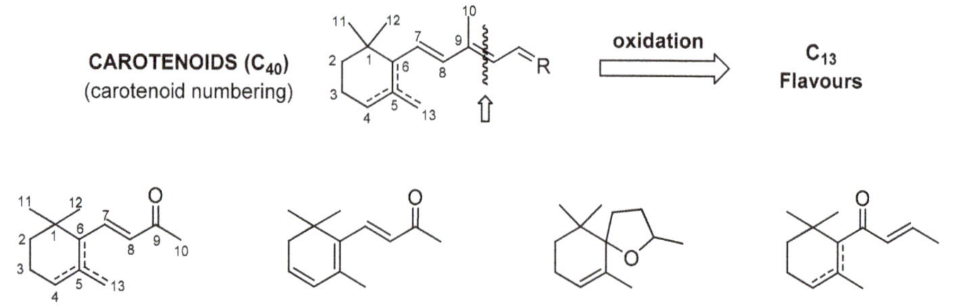

$\Delta_{4,5}$ α-**ionone 1**
$\Delta_{5,6}$ β-**ionone 2**
$\Delta_{5,13}$ γ-**ionone 3**

3,4-dehydro-β-ionone 4

theaspirane 5

$\Delta_{4,5}$ α-**damascone 6**
$\Delta_{5,6}$ β-**damascone 7**

Figure 1. The formation of apocarotenoids through degradation of carotenoids and the seven C_{13} apocarotenoids (compounds **1–7**) selected as substrates for the fungal biotransformation investigated in the present study.

The first consequence of the introduction of a hydroxy- or keto- functional group on these compounds is the decrease of the volatility and the increase of the so-called 'substantivity', namely the long lasting odour of a substance having low vapor pressure. The aforementioned compounds have been recognized as components of different natural flavours. For example 3-hydroxy and 3-keto-α-ionone, 4-hydroxy- and 4-keto-β-ionone and hydroxy-β-damascone isomers have been identified in curry tree [3], eucalyptus honey [4], saffron [5], black tea [6] and tobacco [7,8] respectively, whereas 3-keto-theaspirane (also known under the trade name theaspirone) is the character impact compound of the black tea flavour [9].

All these derivatives are present in vegetables in very minute amounts and the extraction is not a viable process for their production. Consequently, they are currently obtained by chemical synthesis and are not commercially available in their natural form. Since flavours possessing 'natural' status are usually hundreds times as expensive as their synthetic counterparts, any new procedure that provides these compounds in their high value form can be very profitable.

In recent years, some new biocatalytic processes have provided a reliable access route to the most common C_{13} apocarotenoids, such as α- and β-ionone. In addition, the genetic engineering of both carotenoids biosynthesis and carotenoids cleavage pathways in the same microbial host [10] has laid the foundation for the large-scale production of the C_{13} apocarotenoids in natural form.

According to the European and USA legislation, the biotransformation of a natural precursor is a 'natural method' of synthesis [11]. Therefore, we singled out the compounds **1–7** as prospective natural precursors to be used as starting materials for the biotechnological production of different natural flavours.

From a biochemical standpoint, both prokaryotes and eukaryotes are able to degrade the carotenoid frameworks. In spite of this fact, only a limited number of biotransformations of the compounds of type **1–7** have been reported to date. Although the first description of an ionone isomer biotransformation goes back to 1950, when the oxidation of β-ionone in rabbit was investigated [12], only a limited number of studies on this topic took place in the following years. These researches were based mainly on the fungi and bacteria-mediated chemical transformations whereas the exploitation of some specific oxido-reductases were described only recently. In particular, *Aspergillus niger* [13,14], *Lasiodiplodia theobromae* [15], *Cunninghamella blakesleeana* [16], *Botrytis cinerea* [17], *Aspergillus awamori* [18], *Pleurotus sapidus* [19], *Mortierella isabellina* [20] and different *Streptomyces* strains [21] have proved to be active biocatalysts for the transformation of these kind of compounds. Concerning the use of isolated enzymes or the exploitation of a specific enzymatic activities, both cytochrome P450 monooxygenases [22,23] and engineered whole cell biocatalysts expressing mutant P450 monooxygenases [24], were used for the oxidation of different ionone isomers.

It is worth nothing that the biotransformation of some substrates such as γ-ionone, 3,4-dehydro-β-ionone and α-damascone hasn't been investigated yet. This paucity of scientific studies is often due to the difficult availability of some apocarotenoids that can be either substrates or products of the biotransformations. For example, γ-ionone is a rare natural isomer of ionone and can be obtained in high isomeric purity only through demanding multistep syntheses [25–28]. Therefore is not surprising that the fungus-mediated transformation of this compound hasn't been studied yet. In addition, whole cell biotransformations usually afford very complex mixtures of products whose chemical identification, for example by GC or HPLC analysis, require the availability of the corresponding reference standards. The aforementioned compounds are often not commercially available and have to be prepared by specific and multistep chemical syntheses, thus hampering to perform a proper study on the apocarotenoid's biotransformation.

Taking advantage of our previous experience on the stereoselective synthesis of ionone and damascone isomers [2,25–32], we decided to set up a comprehensive study on the biotransformation of the seven natural substrates described above by means of eleven selected fungal species belonging to the three more relevant phylums, namely ascomycota, zygomycota and basidiomycota. More specifically, we selected *Aspergillus niger* and *Mortierella isabellina* because these microorganisms have been already used for the biotransformation of some ionone isomers [13,14,20]. The remaining nine strains were singled out among the plethora of the microorganisms described in the literature based on their prospective biotransformation abilities. In effect, *Nigrospora oryzae*, different *Penicillium* species, *Rhizopus stolonifer*, *Curvularia lunata* and *Fusarium culmorum* have been successful employed in the biotransformation of terpenoids and steroids [33–35] whereas *Geotrichum candidum* was used for the oxidation of the cyclohexanone derivatives [36]. Finally, we selected also the yeasts *Yarrowia lipolytica* and *Xanthophyllomyces dendrorhous* since they are microorganisms of primary interest in the industrial synthesis/degradation of lipids [37] and carotenoids [38], respectively.

The results obtained by our work, beside confirming and extending those described by some previous researches, give new insights on the ability of fungi in the biotransformation of apocarotenoids, establishing their prospective utility for flavour production.

2. Results and Discussion

As described in the introduction, each one of the selected apocarotenoid isomers was incubated with a growing culture of each one of the eleven fungal strains. After a defined period of time (see Experimental Section) the crude biotransformation mixtures were derivatized (by acetylation) and analysed by GC-MS. To this end, several reference standard compounds were prepared by chemical synthesis and then were used for the unambiguous identification of the compounds formed in the biotransformation experiments. In spite of our efforts, a number of these derivatives were not identified. Therefore, in order to spot the most relevant biochemical transformations that each fungal strain is able to perform, we carried out the chromatographic isolation of the unknown metabolites that were formed in relevant amount or that make up of the main part of the transformed derivatives. The structures of these compounds were then elucidated through their comprehensive chemical characterization. More specifically, the combined use of ^{1}H-NMR, ^{13}C-NMR, DEPT experiments, GC-MS and ESI-MS spectroscopy allowed us to identify some compounds that haven't been correlated with any biotransformation experiment yet or that haven't been described in the literature until now.

We first investigated the transformation of the ionone isomers **1–3**. Although only α- and β-ionone are very common in Nature, we deemed that it would be very interesting to also study the reactivity of the rare γ-isomer. In effect, even if the latter three compounds differ only for the position of the cyclohexene double bond, they possess very different reactivity from each other. Therefore, the identification of the derivatives obtained through biotransformation can help in understanding the chemical processes involved in these fungi-mediated reactions.

As collectively described in Figure 2, the main part of the identified metabolites are the result of four different biochemical reactions, namely the oxidation of the methylene functional groups, the reduction of the conjugated double bond, the epoxidation of the 4,5-double bond and the reduction of the keto group. The investigated fungal strains are able to perform both single step reactions and sequential multi-steps transformations, in turn deriving from the combination of the aforementioned four chemical reactions.

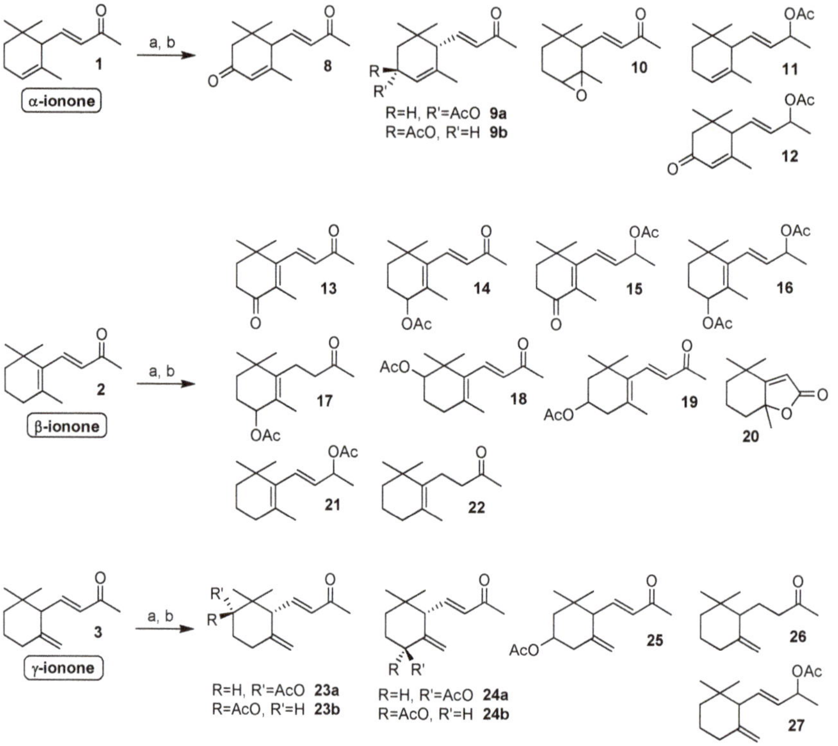

Figure 2. Compounds obtained through biotransformation of α-, β- and γ-ionone by means of the fungal strains employed in the present study. *Reagents and conditions:* (a) fungal strain, malt extract medium (MEM), aerobic conditions, 140 rpm, 20 or 25 °C, 8–20 days; (b) Ac$_2$O/pyridine (Py), 4-dimethylaminopyridine (DMAP) catalyst, room temperature (rt), 6 h.

Overall, ionone isomers **1–3** were converted into derivatives **8–27** in relative amounts described in Table 1. A thorough perusal of these data shows considerable differences in the reactivity of the three isomers. α-Ionone **1** was oxidized almost exclusively at the activated allylic methylene functional group, with the exception of *Geotrichum candidum* that is completely inactive and of *Curvularia lunata* and *Fusarium culmorum* that are also able to oxidize the 4,5-double bond to give the corresponding epoxy-ionone **10**. The relative ratio of the obtained 3-keto-α-ionone **8** and of the diastereoisomeric 3-hydroxy-α-ionones **9a** and **9b** changes significantly based on the fungal strain used. Overall, the microorganisms that better performed the oxidation of **1** were *Aspergillus niger*, *Nigrospora oryzae* and *Fusarium culmorum*.

The global amount of the 3-oxidized metabolites obtained using the latter fungal strains took account of the 74%, 69% and 58% of the crude biotransformation mixture, respectively. On the contrary, none of the tested strains showed notable reductive activity on α-ionone, as confirmed by the modest formation of derivatives **11** and **12**.

Table 1. Results of the fungi-mediated biotransformation of α-, β- and γ-ionone isomers.

Substrate	Biotransformation Products	Fungal Strains and Distribution of the Biotransformation Products [1]										
		A. niger	N. oryzae	G. candidum	Y. lipolytica	P. roqueforti	R. stolonifer	P. corylophilum	M. isabellina	C. lunata	X. dendrorhous	F. culmorum
α-ionone (1)	1	21	2	100	87	82	85	84	62	-	88	24
	8	13	27	-	-	4	2	4	1	2	-	10
	9a	33	7	-	1	3	3	4	9	7	2	21
	9b	28	25	-	1	10	1	6	12	-	1	25
	10	-	-	-	8	-	-	-	-	10	-	3
	11	-	10	-	-	-	-	-	3	-	2	-
	12	-	-	-	-	-	-	-	-	7	-	2
	N.D. [2]	5	31	-	3	1	9	2	13	74	7	15
β-ionone (2)	2	45	15	76	82	87	66	61	84	16	90	40
	13	1	-	5	4	1	3	3	1	-	-	4
	14	39	26	8	1	8	14	16	7	15	3	11
	15	-	-	-	-	-	-	1	-	-	-	-
	16	-	-	-	-	-	-	-	-	23	-	3
	17	-	26	-	-	-	-	-	-	-	-	-
	18	11	-	-	-	-	-	8	-	-	-	-
	19	-	-	-	-	2	-	-	-	-	-	-
	20	-	-	1	-	-	-	-	-	-	-	-
	21	-	2	-	11	-	-	3	1	12	-	3
	22	-	5	-	1	-	-	-	-	-	-	-
	N.D. [2]	4	26	10	1	2	17	8	7	34	7	39
γ-ionone (3)	3	-	6	100	43	83	100	44	85	26	78	21
	23a	35	-	-	-	5	-	14	-	-	-	-
	23b	-	-	-	-	-	-	-	-	-	-	-
	24	2	4	-	-	-	-	-	2	8	-	13
	25	49	2	-	20	-	-	-	-	-	-	10
	26	-	2	-	16	-	-	19	-	-	8	7
	27	-	-	-	-	1	-	6	2	20	5	-
	N.D. [2]	14	86	-	21	11	-	17	11	46	9	49

[1] Percentage of the compound detected by GC-MS analysis of the biotransformation mixture, after extraction and chemical acetylation; [2] N.D. = not determined: the value indicates the overall percentage of the compounds obtained by biotransformation whose chemical structure wasn't assigned.

A more complex outcome were observed when β-ionone **2** was used as substrate. In this case, even if the latter ketone was oxidized mainly at the activated allylic methylene functional group (compounds **13–17**), the biotransformations also afforded the 2-hydroxy-β-ionone derivative **18** (*Aspergillus niger* and *Penicillium corylophilum*) and a little amount of 3-hydroxy-β-ionone derivative **19** (*Penicillium roqueforti*). In addition, the partial oxidation of the ionone side chain is also possible, as shown by the detection of trace (1%) of dihydroactinodiolide **20** in the biotransformation mixture of *Geotrichum candidum*. Interestingly, all the tested fungal strains left unaffected the C(13) methyl group of both α- and β-ionone isomers. In effect, we observed neither the formation of 13-hydroxy-derivatives nor the presence of the epoxy-megastigmaen-9-one isomers that can arise from the intramolecular 1,4-addition of the hydroxy-group to the conjugated double bond [31].

On the contrary, the reduction of the conjugated double bond and of the keto group are chemical transformations of major significance in fungal β-ionone biotransformation. The abovementioned reactions can proceed also on intermediates deriving from the oxidation of the activated position 4 of the β-ionone framework, affording a number of oxidized-reduced derivatives (**15–17**) besides compounds obtained by simple reduction (i.e., **21** and **22**).

Completely different results were observed for the biotransformation of γ-ionone where the oxidation of the allyl methylene group, to afford derivative **24**, appeared to be a path of minor relevance. On the contrary, the oxidation of the positions 2 and 3 of the latter ionone isomer was efficiently performed by *Aspergillus niger* that produced compounds **23a** and **25** in high yield. *Penicillium corylophilum* is also able to convert γ-ionone into the derivative **23a** but for this microorganism, as well as for *Yarrowia lipolytica*, the reductive steps are more relevant than the oxidative ones. In effect, the reduction of the conjugated double bond and of the carbonyl functional group afforded the compound **26** and **27**, respectively. It is worth noting that *Aspergillus niger*, *Penicillium roqueforti* and *Penicillium corylophilum* produced diastereoselectively *cis*-2-hydroxy-γ-ionone (**23a**) and none of the microorganisms tested afforded its diastereoisomer, namely *trans*-2-hydroxy-γ-ionone (**23b**). Differently, the fungi *Nigrospora oryzae*, *Curvularia lunata* and *Fusarium culmorum* transformed efficiently γ-ionone, but the main part of the biotransformation reaction consisted in a mixture of unknown compounds, most likely deriving by the extensive oxidative degradation of the ionone framework.

The second group of the investigated apocarotenoids regards 3,4-dehydro-β-ionone (**4**) and theaspirane (**5**, Figure 3).

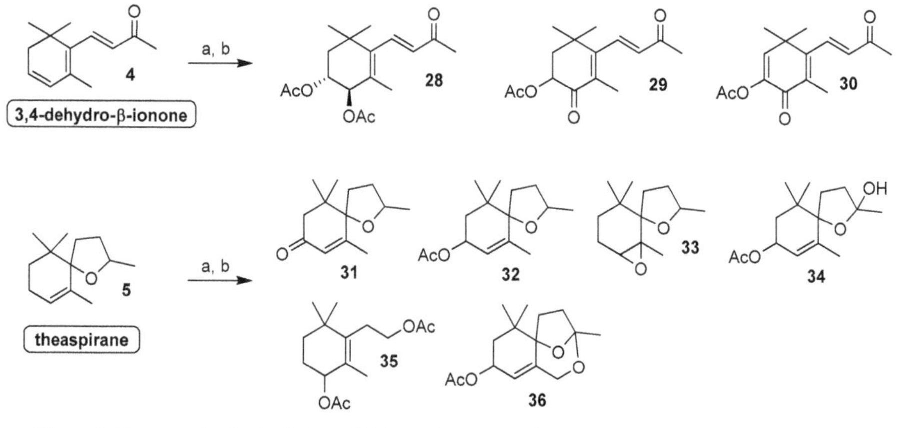

Figure 3. Compounds obtained through biotransformation of 3,4-dehydro-β-ionone and theaspirane by means of the fungal strains employed in the present study. *Reagents and conditions*: (**a**) fungal strain, MEM, aerobic conditions, 140 rpm, 20 or 25 °C, 14–20 days; (**b**) Ac$_2$O/Py, DMAP catalyst, rt, 6 h.

Although these compounds have different chemical structures, they both showed high reactivity and the results of their biotransformation experiments were described together. More specifically, we observed that all the investigated fungal strains were not able to oxidize the allylic positions of compound **4**. We identified as biotransformation products only compounds **28**–**30** (Table 2). These three ionone derivatives originated from the oxidation of the conjugated diene functional group. Most likely the first step is the epoxidation of the 3,4-double bond followed by the opening of the oxirane ring by addition of a molecule of water. This two step mechanism could justify the formation of the compound **28**, possessing *trans* relative configuration of the two contiguous hydroxy groups, as the major diastereoisomer. In addition, the following oxidation of the alcohol functional groups gave rise to the keto derivatives. More specifically, the oxidation at position 4 or at both position 4 and 3 gave compounds **29** and **30**, respectively.

Otherwise, when a spirocyclic ether group replaces the conjugated carbonyl group, our selected fungal strains become able to oxidize the theaspirane framework both at the allylic positions and at the methine carbon linked to the ether oxygen atom. Overall, the main part of the compounds obtained by biotransformation of **5** derive from allylic oxidation. In particular, the experiments performed using *Aspergillus niger*, *Rhizopus stolonifer* and *Fusarium culmorum* showed a total content of the compounds **31** and **32** that accounts for at least 60% of the reaction mixtures. This result is noteworthy as theaspirone **31** is a relevant natural flavour and the corresponding alcohol (**32** is the acetylated derivative) can be regarded as its direct precursor. In effect, the preparation of **31** in natural form could be possible by means of oxidation of the allyl alcohol, for example using a biocatalytic transformation involving alcohol dehydrogenases.

Jointly with allylic oxidation, we observed also the oxidation of the ether functional group. Hence, the latter moiety was transformed into the hemiketal and ketal groups as well in a completely rearranged framework, as demonstrated by the isolation and characterization of compounds **34**, **36** and **35**, respectively. Most likely, the latter derivatives are the result of a multistep oxidation reaction. The transformation of the allyl methylene and methyl groups proceeds faster than the oxidation of ether functional group, which is finally converted into a hemiketal functional group. Accordingly, *Nigrospora oryzae* and *Mortierella isabellina* completely oxidized theaspirane (**5**) to give two biotransformation mixtures containing 28% and 50% of compound **34**, respectively. It is worth mentioning that the latter compound is the direct precursor of 8,9-dehydrotheaspirone [32], a relevant apocarotenoid flavour identified in white-fleshed nectarines [39].

The biotransformation experiments performed using *Curvularia lunata* gave results of more complex interpretation affording a plethora (60% of the mixture) of undetermined compounds beside the 3%, 8%, 12% and 9% of derivatives **32**, **34**, **35** and **36**, respectively. Compound **35** could result from a multistep transformation comprising both of oxidation and transposition reactions. The C(9) oxidation is responsible for the formation of the hemiketal functional group that is in equilibrium with the less stable open hydroxy-ketone form. The Baeyer-Villiger oxidation of the latter ketone functional group could explain the formation of the primary alcohol acetate moiety whereas the 1,3-allyl transposition of the tertiary hydroxy group could take account of the unexpected position of the C(4) acetate group in compound **35**. Similarly, the formation of compound **36** in *Curvularia lunata*–mediated biotransformation could be explained by the hydroxylation of the carbons placed in position 3, 9 and 13 of the ionone framework.

The subsequent reaction of the obtained primary alcohol group with the hemiketal functional group should afford the more thermodynamically stable tricyclic ketal **36**. It is worth nothing that a natural compound having an identical ketal structure, but devoid of the hydroxy group at the C(4), is an aroma component of quince brandy [40]. The comparison of the NMR data measured for **36** with those reported for the natural flavour, allowed us to assign the above described chemical structure suggesting a new synthetic approach to this chemical framework by fungal biotransformation of theaspirane.

Table 2. Results of the fungi-mediated biotransformation of 3,4-dehydro-β-ionone and theaspirane.

Substrate	Biotransformation Products	Fungal Strains and Distribution of the Biotransformation Products [1]										
		A. niger	N. oryzae	G. candidum	Y. lipolytica	P. roqueforti	R. stolonifer	P. corylophilum	M. isabellina	C. lunata	X. dendrorhous	F. culmorum
3,4-dehydro-β-ionone (4)	4	48	-	45	87	45	80	76	71	52	6	-
	28	25	69	18	-	32	7	22	8	26	26	35
	29	3	2	17	1	9	3	-	7	8	22	20
	30	2	3	12	1	6	2	-	5	6	21	17
	N.D. [2]	22	26	8	3	8	8	2	9	8	25	28
theaspirane (5)	5	20	-	56	82	31	-	22	-	8	46	-
	31	9	12	11	4	19	5	16	-	-	3	28
	32	51	25	15	1	28	62	34	-	3	28	34
	33	-	-	-	-	2	-	-	-	-	2	2
	34	-	28	-	-	-	6	-	50	8	-	9
	35	-	-	-	-	-	-	-	-	12	-	-
	36	-	-	-	-	-	-	-	-	9	-	-
	N.D. [2]	20	35	18	1	20	27	28	50	60	21	27

[1] Percentage of the compound detected by GC-MS analysis of the biotransformation mixture, after extraction and chemical acetylation; [2] N.D. = not determined: the value indicates the overall percentage of the compounds obtained by biotransformation whose chemical structure wasn't assigned.

A completely different reactivity was observed in the biotransformation of the α- and β-damascone isomers **6** and **7** (Figure 4). The latter apocarotenoids are isomers of α- and β-ionone respectively, as each one damascone isomer is interconvertible into the corresponding ionone isomer by 1,3-shift of the enone moiety.

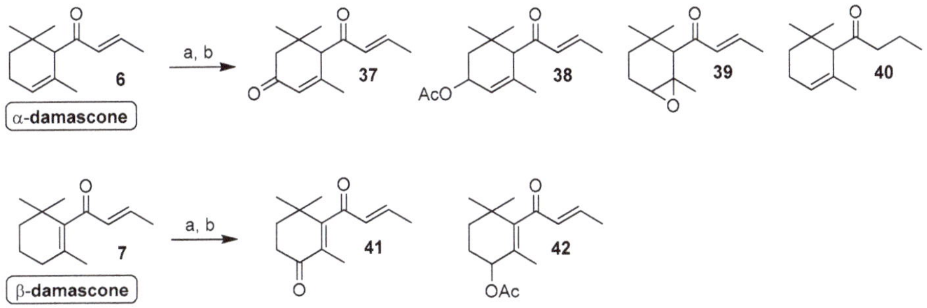

Figure 4. Compounds obtained through biotransformation of α- and β-damascone by means of the fungal strains employed in the present study. *Reagents and conditions*: (**a**) fungal strain, MEM, aerobic conditions, 140 rpm, 20–25 °C, 14–20 days; (**b**) Ac$_2$O/Py, DMAP catalyst, rt, 6 h.

Despite the structural similarity of these compounds, none of the investigated fungal strains was able to reduce the carbonyl functional group present in the damascone framework. In addition, the fungi-mediated oxidative transformations of damascone isomers are restricted mainly to the allyl methylene functional groups, as indicated by the formation of derivative **37** and **38** from α-damascone and derivatives **41** and **42** from β-damascone (Table 3). The other position of the damascone framework were unaffected with the exception of the 4,5-double bond that was oxidized by both *Mortierella isabellina* and *Xanthophyllomyces dendrorhous* to produce a very minor amount of the epoxy-α-damascone **39**. Similarly, we did not record any reductive reactions with the exception of the conjugated double bond of the α-damascone isomer that was reduced by *Penicillium corylophilum* to give a small amount of compound **40**.

Overall, using the described fungal strains, we observed that the damascone isomers are less reactive than the corresponding ionone isomers. This effect is more pronounced for the β-isomer where only *Nigrospora oryzae* and *Fusarium culmorum* produced a significant amount of the corresponding 4-hydroxydamascone, identified as acetyl derivative **42** (23 and 22%, respectively).

Our results seem in contrast to those reported in a recent study on the fungal biotransformation of β-damascone [20], where a different *Mortierella isabellina* strain provides 4-hydroxydamascone in a much higher yield. The recorded differences between these experimental data could be justified considering both the substrate concentrations and the different biocatalytic activity among fungal strains belonging to the same species.

In particular the substrate concentration seem to be the most relevant factor as apocarotenoid derivatives show significant toxicity for many fungal strains and high concentration of these compounds could inhibit their growth as well as their biocatalytic activity. We performed all the investigated biotransformation experiments using a substrate concentration of about 2.5 g/L whereas, in the above-mentioned work, the β-damascone concentration was set to 0.1 g/L. Our choice is justified by the need of devising a preparative protocol for fungi-mediated ionone biotransformation. Since the synthesis of this kind of flavours, in natural form, can show industrial significance only working with substrate concentrations superior to 1 g/L, we set the above indicated concentration for all experiments. As a consequence, it is reasonable that our selected fungal strains could have transformed better both damascone as well as ionone and theaspirane isomers if they have been used in lower concentration.

Table 3. Results of the fungi-mediated biotransformation of α- and β-damascone isomers.

Substrate	Biotransformation Products	Fungal Strains and Distribution of the Biotransformation Products [1]									
		A. niger	N. oryzae	G. candidum	Y. lipolytica	P. roqueforti	R. stolonifer	P. corylophilum	M. isabellina	C. lunata	X. dendrorhous	F. culmorum
α-damascone (6)	6	56	89	55	93	100	100	24	30	96	39	56
	37	20	5	10	-	-	-	-	22	-	10	5
	38	5	2	10	-	-	-	14	17	3	21	30
	39	-	-	-	-	-	-	-	2	-	3	-
	40	-	-	-	-	-	-	13	-	-	-	-
	N.D. [2]	19	4	25	7	-	-	49	29	1	27	9
β-damascone (7)	7	98	70	95	95	100	88	95	95	92	95	57
	41	-	2	-	-	-	1	-	-	-	-	4
	42	1	23	1	1	-	2	-	2	7	1	22
	N.D. [2]	1	5	4	4	-	9	5	3	1	4	17

[1] Percentage of the compound detected by GC-MS analysis of the biotransformation mixture, after extraction and chemical acetylation; [2] N.D. = not determined: the value indicates the overall percentage of the compounds obtained by biotransformation whose chemical structure wasn't assigned.

3. Materials and Methods

3.1. Materials and General Methods

All air and moisture sensitive reactions were carried out using dry solvents and under a static atmosphere of nitrogen. All solvents and reagents were of commercial quality and were purchased from Sigma-Aldrich (St. Louis, MO, USA). A large number of reference standard compounds were synthesized in our laboratory and were used for the unambiguous identification of the compounds formed in the biotransformation experiments. α-Ionone, γ-ionone and α-damascone were used in racemic form. Commercial theaspirane consists of an equimolar mixture of racemic diastereoisomers. γ-Ionone (**3**) and 3,4-dehydro-β-ionone (**4**) were prepared starting from α-ionone, according to the procedures previously described by us [26–28].

The keto derivatives: 3-keto-α-ionone (**8**), 4-keto-β-ionone (**13**), 3-keto-α-damascone (**37**), 4-keto-β-damascone (**41**), theaspirone (**31**), 3-keto-α-ionol acetate (**12**) and 4-keto-β-ionol acetate (**15**) were prepared by oxidation of α-ionone, β-ionone, α-damascone, β-damascone, theaspirane, α-ionol acetate (**11**) and β-ionol acetate (**21**), respectively. The oxidation reactions were performed using TBHP/MnO_2 as oxidant according to our previously reported procedure [41].

The diastereoisomeric forms of 4,5-epoxy-α-ionone (**10**), 4,5-epoxy-α-damascone (**39**), 4,5-epoxy-theaspirane (**33**) as well as 5,6-epoxy-β-ionone and 5,6-epoxy-β-damascone were prepared by epoxidation of α-ionone, α-damascone, theaspirane, β-ionone and β-damascone, respectively, using *m*-chloroperbenzoic acid and CH_2Cl_2 as solvent.

α-7,8-Dihydroionones, β-7,8-dihydroionone (**22**) and γ-7,8-dihydroionone (**26**) were prepared by reduction of α-, β- and γ-ionone respectively, using hydrogen and Ni Raney as catalyst for α-ionone [42] and Bu_3SnH and $(Ph_3P)_2PdCl_2$ as catalyst for β- and γ-ionone [25,26]. α-8,9-Dihydrodamascone (**40**) was prepared by reduction of α-damascone using $NaBH_4$ in methanol. β-8,9-Dihydrodamascone was prepared by addition of propylmagnesium bromide to β-cyclocitral followed by oxidation of the resulting carbinol using Dess-Martin periodinane [43].

β-Ionol acetate (**21**, racemic), α-ionol acetate (**11**, as a mixture of two racemic diastereoisomers), γ-ionol acetates (**27**, as a mixture of two racemic diastereoisomers) and 3-acetoxy-theaspirane (**32**) (as a mixture of four racemic diastereoisomers) were prepared by chemical acetylation (Ac_2O/Py) of the corresponding alcohols, which were in turn obtained through the reduction of α-, β-, γ-ionone and theaspirone, respectively, using $NaBH_4$ in methanol.

cis-2-Acetoxy-α-ionone, and *cis*-2-acetoxy-γ-ionone (**23a**) were prepared starting from 2,8,8-trimethyl-6-oxabicyclo[3.2.1]oct-2-en-7-one and 8,8-dimethyl-2-methylene-6-oxabicyclo[3.2.1]octan-7-one (kaharana lactone) respectively, according to the synthetic procedure developed by Audran [44]. The latter lactones were in turn prepared from racemic *cis*-2-hydroxy-α-cyclogeraniol and *cis*-2-hydroxy-γ-cyclogeraniol [45] by oxidation using BAIB and TEMPO as catalyst. In addition, the partial reduction of the two isomeric lactones afforded the corresponding lactols, whose condensation with acetone [46] followed by acetylation (Ac_2O/Py) of the crude reaction mixtures, gave the *cis*/*trans* mixtures of acetoxy-α-ionone and acetoxy-γ-ionone, respectively, that were used as GC-MS reference standards for the identification of the corresponding *trans* isomers.

Racemic 2-acetoxy-β-ionone (**18**) was prepared starting from 2-hydroxy-β-cyclogeraniol [45] by selective oxidation of the primary alcohol functional group by MnO_2 in $CHCl_3$, condensation with acetone [46] and acetylation (Ac_2O/Py) of the obtained hydroxy-ionone derivative.

Samples of 3-acetoxy-α-ionone (**9**) (2:1 *cis*/*trans* mixture), of 4-acetoxy-γ-ionone (**24**) (4:1 *cis*/*trans* mixture) and of 3-acetoxy-β-ionone (**19**) were prepared starting from α-ionone according to the procedure described by Tu [47], by Serra [42] and Khachik [48], respectively. A sample of 3-acetoxy-α-damascone (**38**) (1:1 *cis*/*trans* mixture) was prepared starting from ethyl 3-hydroxy-α-cyclogeraniate according to the procedure described by Takei [49].

4-Acetoxy-β-ionone (**14**) and 4-acetoxy-β-damascone (**42**) were prepared starting from 4,5-epoxy-α-ionone and 4,5-epoxy-α-damascone, respectively, by means of NaOMe mediated transposition followed by chemical acetylation (Ac$_2$O/Py) of the obtained allyl alcohols. A sample of 4-acetoxy-β-ionol acetate (**16**) (1:1 mixture of diastereoisomers) was prepared from 4-keto-β-ionol acetate by reduction with NaBH$_4$ in methanol followed by chemical acetylation (Ac$_2$O/Py).

A sample of 4-acetoxy-β-7,8-dihydroionone acetate (**17**) was prepared from 4-hydroxy-β-ionone by reduction with Ph$_3$SiH followed by chemical acetylation (Ac$_2$O/Py), according to the procedure described by Pascual [50].

3,4-Diacetoxy-β-ionone (**28**) (cis/trans mixture), was prepared from 3,4-dehydro-β-ionone according to the procedure described by Buschor [51]. The oxidation of 4-keto-β-ionone with IBDA in methanol [52] afforded 3-hydroxy-4-keto-β-ionone that was further oxidized using oxygen in presence of tBuOK [53] to give 3,4-diketo-β-ionone. The acetylation (Ac$_2$O/Py) of the latter two compounds afforded 3-acetoxy-4-keto-β-ionone (**29**) and 3-acetoxy-4-keto-2,3-dehydro-β-ionone (**30**). Racemic dihydroactinodiolide (**20**) and 2-hydroxy-2,6,10,10-tetramethyl-1-oxaspiro[4.5]dec-6-en-8-yl acetate (**34**) were prepared as described previously [32,54].

A comprehensive characterization of the above described reference standards is reported in the Supplementary Materials section.

3.2. Analytical Methods and Characterization of the Products Deriving from the Biotransformation Experiments

The crude biotransformation mixtures obtained according to the procedures described below were then acetylated by treatment with pyridine/acetic anhydride (2 mL of a 2:1 mixture) and catalytic DMAP (10 mg) for 6 hours at rt. The obtained acetylated mixture was analyzed by GC-MS. The compounds whose chemical structure couldn't be assigned only by GC-MS analysis were isolated from the biotransformation mixtures by means of chromatographic separation and then characterized by NMR analysis and GC-MS or ESI-MS analysis.

^{1}H- and ^{13}C-NMR spectra and DEPT experiments were recorded/performed at 400, 100 and 100 MHz, respectively, in CDCl$_3$ solutions at rt using an AC-400 spectrometer (Bruker, Billerica, MA, USA); ^{13}C spectra are proton decoupled; chemical shifts in ppm rel to internal SiMe$_4$ (=0 ppm).

TLC: silica gel 60 F_{254} plates (Merck, Kenilworth, NJ, USA). Column chromatography: silica gel.

Melting points were measured on a Reichert apparatus, equipped with a Reichert microscope, and are uncorrected.

Mass spectrum were recorded on a ESQUIRE 3000 PLUS spectrometer equipped with an ESI detector (Bruker, Billerica, MA, USA) or by GC-MS analyses.

GC-MS analyses: *HP-6890* gas chromatograph equipped with a 5973 mass detector, using a HP-5MS column (30 m × 0.25 mm, 0.25 μm film thickness; Hewlett Packard, Palo Alto, CA, USA) with the following temp. program: 60° (1 min)—6°/min—150° (1 min)—12°/min—280° (5 min); carrier gas, He; constant flow 1 mL/min; split ratio, 1/30; t_R given in min: t_R(**1**) 16.23, t_R(**2**) 17.57, t_R(**3**) 16.53, t_R(**4**) 17.53, t_R(**5**) 13.40 and 13.77, t_R(**6**) 15.44, t_R(**7**) 15.91, t_R(**8**) 20.42, t_R(**9a**) 21.41, t_R(**9b**) 21.61, t_R(**10**) 18.63 and 18.72, t_R(**11**) 17.78, t_R(**12**) 21.52, t_R(**13**) 20.61, t_R(**14**) 21.82, t_R(**15**) 22.03, t_R(**16**) 22.56, t_R(**17**) 21.59, t_R(**18**) 22.18, t_R(**19**) 22.23, t_R(**20**) 18.61, t_R(**21**) 18.55, t_R(**22**) 16.51, t_R(**23a**) 21.51, t_R(**24a**) 21.32, t_R(**24b**) 21.60, t_R(**25**) 21.79, t_R(**26**) 15.67, t_R(**27**) 17.47, 17.58, 18.01 and 18.08, t_R(**28**) 24.10, t_R(**29**) 23.75, t_R(**30**) 23.92, t_R(**31**) 19.02 and 19.19, t_R(**32**) 20.17, 20.33, 20.50 and 20.61, t_R(**33**) 15.14 and 15.50, t_R(**34**) 19.37 and 19.67, t_R(**35**) 21.78, t_R(**36**) 21.64, t_R(**37**) 20.15, t_R(**38**) 21.17 and 21.30, t_R(**39**) 17.84, t_R(**40**) 14.62, t_R(**41**) 20.91, t_R(**42**) 20.93, t_R(7,8-dihydro-α-ionone) 15.95, t_R(7,8-dihydro-β-damascone) 20.95, t_R(cis-2-acetoxy-α-ionone) 21.42, t_R(trans-2-acetoxy-α-ionone) 21.36.

3.3. Microorganisms and Biotransformation Experiments

Geotrichum candidum (DSM 10452), Yarrowia lipolytica (DSM 8218), Rhizopus stolonifer (DSM 855), Xanthophyllomyces dendrorhous (DMS 5626), Curvularia lunata (CBS 215.54), Mortierella isabellina (CBS 167.60), Aspergillus niger (CBS 626.26) were purchased from the DSMZ (Braunschweig, Germany) or CBS-KNAW (Utrecht, The Netherlands) collections.

Penicillium corylophilum (MUT 5838), *Nigrospora oryzae* (MUT 5844), *Penicillium roqueforti* (MUT 5856) and *Fusarium culmorum* (MUT 5855) were isolated as axenic cultures in our laboratory, then identified by the Mycotheca Universitatis Taurinensis (MUT) of the University of Turin and finally deposited in the same institution under the collection number given in brackets.

All the biotransformations were carried out in triplicate and the presented results are the average of three experimental runs.

3.3.1. Representative Procedures for Biotransformations

The experimental conditions used for the biotransformations are based on the type of microorganism used. Here is described a general procedure depending on the different morphological features regarding the various active grow mycelia. The main ones could be classified in yeast-shape mycelia (*Xanthophyllomyces dendrorhous*, *Geotrichum candidum* and *Yarrowia lipolytica*) and spore-forming mycelia (*Aspergillus niger*, *Rhizopus stolonifer*, *Curvularia lunata*, *Penicillium corylophilum*, *Nigrospora oryzae*, *Penicillium roqueforti*, *Mortierella isabellina* and *Fusarium culmorum*). In the first case, a small amount of the active mycelia grew previously in a petri dish, was suspend in 1 mL of sterile water and then inoculated in a 100 mL conical Pyrex flask containing 40 mL of Malt Extract Medium (MEM) for 2 days at 25 °C and 140 rpm (with exception of *Xanthophyllomyces dendrorhous* that was grown at 20 °C). After this period, the cells were centrifuged 3 minutes, (rt, 3220·g) and collected removing the media. The cells (approx. 600 mg wet-weight) were suspended in 3 mL of sterile water than 350 µL of the same suspension were used for inoculating each biotransformation flask containing 40 mL of MEM. In order to ensure aerobic conditions, the flasks were sealed with cellulose plugs. The microorganism was leave to growth for 2 days and then was treated with a solution of 100 mg of substrate dissolved in 400 µL of DMSO. Generally, after 14 days from the substrate injection using the growing condition described above, the reaction media was filtered under vacuum through a celite pad then was extracted 3 times with ethyl acetate. The organic phase was separated, dried on Na_2SO_4 and the solvent removed at reduced pressure to give the crude biotransformation mixture.

In the case of the spore-forming mycelia, the spore were collected from a sporulated surface cultures and suspended in 3 mL of sterile water. After that, 350 µL of the same suspension were used for inoculating each biotransformation flask containing 40 mL of MEM. The subsequent steps are the same of that described previously. The unique exception was carried out for *Penicillium corylophilum*, in the case of γ-ionone. The toxicity of the compound forced us to keep its concentration lower than the others (7.5 mM) and to block the biotransformation earlier (8 days). After this period, the most important products are degraded. In the case of *Fusarium culmorum* the biotransformation was blocked after 20 days instead of 14 days because the activity of the fungus did not stop in the prefixed time.

3.3.2. Preparative Biotransformations and Chemical Characterization of Compounds **23a, 25, 28, 35** and **36**

For the main part of the strains tested, the GC-MS analysis of the crude biotransformation mixtures indicated the presence of different compounds whose chemical structures could not be assigned only on the basis of our reference standards. The unidentified peaks taking account of less than 5% of the overall percentage of the compounds obtained by biotransformation were collectively indicated as 'not determined'. Otherwise, compounds **23a, 25, 28, 35** and **36** were isolated from the fermentation broths by extraction and chromatographic separation and then submitted to chemical characterization. Different reasons prompted us to undertake the isolation procedure. First of all, these compounds were a relevant part of the biotransformation mixture and their MS fragmentations clearly indicated a chemical structure deriving from the corresponding starting materials. The compounds **25** was not one of the references standard available from our laboratory. Hence, we identified this compound only after its isolation and chemical characterization. Both the diastereoisomeric forms of the compound **25** have been described in the literature [55] but only low resolution ^{1}H-NMR data was reported. Therefore, we were not able to assign the relative configuration to the compound obtained by biotransformation. Differently, compound **36** is completely new and its analytic data has not described yet. Concerning compound **28**, we observed that its diastereoisomeric forms (*trans* and *cis* isomers) have the same retention time by GC-MS analysis. As a consequence, the isolation of compound **28** from the biotransformation mixture followed by its NMR analysis was necessary in order to understand what was the main isomer formed. Finally, the case of compound **35** is singular. It is the only compound obtained by biotransformation that was formed through a Baeyer-Villiger oxidation. Consequently, acetate **35** was completely unexpected and the proper reference standard was not synthesized.

Hereafter we reported the procedure for the preparative biotransformation experiments allowing the isolation of compounds **23a, 25, 28, 35** and **36** as well as their main analytic data.

According to the procedure described before for the preparation of the inoculum of spore forming mycelia, *Aspergillus niger*, *Nigrospora oryzae* and *Curvularia lunata*, were inoculated in three 1 L conical pyrex flasks containing 400 mL of MEM. The microorganisms were left to grow at 25 °C and 140 rpm for 2 days. Hence the cultures of *Aspergillus niger*, *Nigrospora oryzae* and *Curvularia lunata*, were treated with a solution of 1 g of γ-ionone, 3,4-dehydro-β-ionone and theaspirane, respectively, each one dissolved in 3 mL of DMSO. After 14 days from the substrate injection, using the growing condition described above, the reaction media was filtered through a celite pad, the filter was washed with ethyl acetate and the filtrate was extracted 3 times with the same solvent. The combined organic phases were separated, were washed with brine, dried on Na_2SO_4 and the solvent was removed under reduced pressure. The residue was then acetylated by treatment with pyridine/acetic anhydride (10 mL of a 2:1 mixture) and catalytic DMAP (10 mg) for 6 h at rt. The acetylating mixture (Py/Ac_2O) was then removed under reduced pressure and the resulting oil was purified by chromatography using *n*-hexane/AcOEt mixture as eluent.

The biotransformation of γ-ionone performed using *Aspergillus niger* allowed isolating 0.21 g (18% yield) of compound **23a** and 0.36 g (28% yield) of compound **25** as a single diastereoisomeric form (configuration not determined):

cis-2-Hydroxy-γ-ionone acetate (**23a**) = (1*SR*,3*RS*)-2,2-Dimethyl-4-methylene-3-((*E*)-3-oxobut-1-en-1-yl)cyclohexyl acetate. ^{1}H-NMR: δ = 6.97 (dd, *J* = 15.8, 9.9 Hz, 1H), 6.10 (d, *J* = 15.8 Hz, 1H), 4.88 (s, 1H), 4.74 (dd, *J* = 9.2, 4.0 Hz, 1H), 4.61 (s, 1H), 2.65 (d, *J* = 9.9 Hz, 1H), 2.41 (dt, *J* = 14.0, 5.4 Hz, 1H), 2.37–2.05 (m, 1H) 2.28 (s, 3H), 2.07 (s, 3H), 1.93–1.83 (m, 1H), 1.72–1.57 (m, 1H), 0.91 (s, 6H). ^{13}C-NMR δ = 197.9 (C), 170.4 (C), 146.0 (C), 145.5 (CH), 132.9 (CH) 111.1 (CH_2), 77.8 (CH), 55.9 (CH), 39.0 (C), 31.2 (CH_2), 27.6 (CH_2), 27.3 (Me), 26.2 (Me), 21.1 (Me), 17.8 (Me). GC-MS (EI): *m/z* (%) = 250 [M^+] (12), 235 [M^+ − Me] (1), 208 (13), 190 (45), 175 (40), 165 (36), 147 (100), 131 (23), 122 (39), 109 (96), 91 (34), 79 (35), 71 (12).

3-Hydroxy-γ-ionone acetate (**25**) = (*E*)-3,3-Dimethyl-5-methylene-4-(3-oxobut-1-en-1-yl)cyclohexyl acetate. ^{1}H-NMR: δ = 6.82 (dd, *J* = 15.8, 10.1 Hz, 1H), 6.13 (d, *J* = 15.8 Hz, 1H), 4.97 (s, 1H), 4.95–4.86 (m, 1H), 4.63 (s, 1H), 2,73 (dd, *J* = 12.6, 4.9 Hz, 1H), 2.58 (d, *J* = 10.1 Hz, 1H), 2.29 (s, 3H), 2.15–2.01 (m, 1H), 2.03 (s, 3H), 1.85 (dd, *J* = 12.6, 4.5 Hz, 1H), 1.45 (t, *J* = 12.1 Hz, 1H), 0.95 (s, 3H), 0.92 (s, 3H). ^{13}C-NMR: δ = 197.8 (C), 170.3 (C), 145.3 (CH), 144.5 (C), 134.0 (CH), 112.5 (CH_2), 69.8 (CH), 56.0 (CH), 45.5 (CH_2), 41.1 (CH_2), 35.8 (C), 30.4 (Me), 27.3 (Me), 21.4 (Me), 21.3 (Me). GC-MS (EI): *m/z* (%) = 250 [M^+] (<1), 235 [M^+ − Me] (<1), 190 (28), 175 (29), 157 (14), 147 (100), 131 (32), 119 (20), 105 (39), 91 (27), 79 (15), 69 (14), 55 (8).

The biotransformation of 3,4-dehydro-β-ionone performed using *Nigrospora oryzae* allowed isolating 0.56 g (34% yield) of compound **28** as a 5:1 mixture of *trans*/*cis* isomers:

trans-3,4-Dihydroxy-β-ionone diacetate (**28**) = (1*RS*,2*RS*)-3,5,5-Trimethyl-4-((*E*)-3-oxobut-1-en-1-yl)cyclohex-3-ene-1,2-diyl diacetate. ^{1}H-NMR: δ = 7.10 (dq, *J* = 16.4, 0.9 Hz, 1H), 6.11 (d, *J* = 16.4 Hz, 1H), 5.51 (d, *J* = 7.8 Hz, 1H), 5.17–5.09 (m, 1H), 2.28 (s, 3H), 2.06 (s, 3H), 2.00 (s, 3H), 1.88–1.66 (m, 2H), 1.62 (br s, 3H), 1.17 (s, 3H), 1.07 (s, 3H). ^{13}C-NMR: δ = 197.8 (C), 170.7 (C), 170.4 (C), 141.0 (CH), 140.2 (C), 133.7 (CH),

128.5 (C), 74.7 (CH), 70.3 (CH), 41.4 (CH$_2$), 36.4 (C), 29.7 (Me), 27.7 (Me), 27.5 (Me), 21.1 (Me), 20.8 (Me), 16.5 (Me). MS (ESI): 331.2 (M$^+$ + Na).

The biotransformation of theaspirane performed using *Curvularia lunata* allowed isolating 95 mg (7% yield) of compound **35** and 65 mg (5% yield) of compound **36**:

2-(3-Acetoxy-2,6,6-trimethylcyclohex-1-en-1-yl)ethyl acetate (**35**). ^{1}H-NMR: δ = 5.13 (t, *J* = 4.6 Hz, 1H), 4.05 (dd, *J* = 9.4, 7.5 Hz, 2H), 2.46–2.37 (m, 2H), 2.06 (br s, 6H), 1.92–1.79 (m, 1H), 1.74–1.52 (m, 2H), 1.66 (s, 3H), 1.44–1.33 (m, 1H), 1.08 (s, 3H), 1.00 (s, 3H). ^{13}C-NMR: δ = 171.0 (C), 171.0 (C), 139.5 (C), 128.2 (C), 72.3 (CH), 63.3 (CH$_2$), 35.1 (C), 34.6 (CH$_2$), 28.2 (Me), 28.0 (CH$_2$), 26.8 (Me), 25.3 (CH$_2$), 21.3 (Me), 21.0 (Me), 16.8 (Me). GC-MS (EI): *m*/*z* (%) = 268 [M$^+$] (<1), 226 (13), 208 [M$^+$ − AcOH] (16), 166 (40), 148 (17), 133 (100), 120 (28), 110 (36), 91 (16), 79 (8).

3,6,6-Trimethyl-1,4,5,6,7,8-hexahydro-3H-3,5a-epoxybenzo[c]oxepin-8-yl acetate (**36**). ^{1}H-NMR: δ = 5.43–5.36 (m, 1H), 5.31 (s, 1H), 4.49 (dt, *J* = 14.2, 2.1 Hz, 1H), 4.20 (d, *J* = 14.2 Hz, 1H), 2.30–1.40 (m, 6H), 2.02 (s, 3H), 1.47 (s, 3H), 1.11 (s, 3H), 1.01 (s, 3H). ^{13}C-NMR: δ = 170.5 (C), 138.7 (C), 116.3 (CH), 105.3 (C), 86.0 (C), 68.4 (CH), 63.7 (CH$_2$), 40.7 (CH$_2$), 35.3 (C), 34.3 (CH$_2$), 30.5 (CH$_2$), 24.6 (Me), 24.4 (Me), 23.3 (Me), 21.2 (Me). GC-MS (EI): *m*/*z* (%) = 266 [M$^+$] (1), 236 (50), 224 (8), 207 [M$^+$ − AcO] (59), 195 (36), 178 (27), 164 (84), 153 (100), 136 (61), 121 (59), 107 (62), 91 (53), 79 (29).

4. Conclusions

Our work provides some relevant findings. First, we demonstrated that fungi are able to perform different biotransformations on the isomeric forms of the C$_{13}$ apocarotenoids ionone, theaspirane and damascone. With respect to the eleven strains tested, we observed that the most common chemical transformations are oxidation reactions that afford oxygenated products such as hydroxy- keto- or epoxy-derivatives. On the contrary, the reduction of the keto groups or the reduction of the double bond functional groups are less relevant transformations, occurring for few substrates and yielding a minority amount of products.

A very significant feature of our study concern the prospective applicability of the fungi-mediated biotransformation of apocarotenoids for the synthesis of high value natural flavours. Since some ionone, damascone and theaspirane isomers are available in natural form and the biotransformation of a natural precursor is considered a 'natural method' of synthesis, the flavours obtained by means of the fungi-mediated reactions described above possess the natural status and could be commercialized accordingly.

Finally, we would like to highlight that our microbial biotransformations allow the preparation of many derivatives whose synthesis, using the classical chemical reactions, is very difficult. For example, different fungal strains proved to be able to oxidize some inactivated positions of the ionone or theaspirane framework, such as the position 2 and 3 of the β- and γ-ionone and the methine carbon linked to the theaspirane oxygen atom. By means of these microbial capabilities, we isolated one new compound (compound **36**) and we devised a new biocatalytic procedure for the synthesis of 2-hydroxy-γ-ionone, 3-hydroxy-γ-ionone, 3,4-dihydroxy-β-ionone and 2,6,10,10-tetramethyl-1-oxa-spiro[4.5]dec-6-ene-2,8-diol (identified as its monoacetate **34**), which are natural apocarotenoids or their direct precursors.

Supplementary Materials: The Supplementary Materials are available online.

Author Contributions: S.S. and D.D.S. equally contributed to the conceptualization of this study. S.S. and D.D.S. equally contributed to conceive, design and perform the experiments as well as to analyze the data. S.S. wrote the paper.

Funding: This research was funded by [Cariplo Fundation] grant number [2014-0568 INBOX (Innovative Biocatalytic Oxidations)].

Acknowledgments: The authors thank Cariplo Foundation for supporting this study.

Conflicts of Interest: The authors declare no conflict of interest.

References

1. Walter, M.H.; Strack, D. Carotenoids and their cleavage products: Biosynthesis and functions. *Nat. Prod. Rep.* **2011**, *28*, 663–692. [CrossRef] [PubMed]
2. Serra, S. Recent advances in the synthesis of carotenoid-derived flavours and fragrances. *Molecules* **2015**, *20*, 12817–12840. [CrossRef] [PubMed]
3. Ma, Q.-G.; Wang, Y.-G.; Liu, W.-M.; Wei, R.-R.; Yang, J.-B.; Wang, A.-G.; Ji, T.-F.; Tian, J.; Su, Y.-L. Hepatoprotective sesquiterpenes and rutinosides from *Murraya koenigii* (L.) Spreng. *J. Agric. Food Chem.* **2014**, *62*, 4145–4151. [CrossRef] [PubMed]
4. Schievano, E.; Morelato, E.; Facchin, C.; Mammi, S. Characterization of markers of botanical origin and other compounds extracted from unifloral honeys. *J. Agric. Food Chem.* **2013**, *61*, 1747–1755. [CrossRef] [PubMed]
5. Li, C.-Y.; Wu, T.-S. Constituents of the stigmas of *Crocus sativus* and their tyrosinase inhibitory activity. *J. Nat. Prod.* **2002**, *65*, 1452–1456. [CrossRef] [PubMed]
6. Ina, K.; Etō, H. 3-keto-β-ionone in the essential oil from black tea. *Agric. Biol. Chem.* **1971**, *35*, 962–963. [CrossRef]
7. Fujimori, T.; Kasuga, R.; Matsushita, H.; Kaneko, H.; Noguchi, M. Neutral aroma constituents in burley tobacco. *Agric. Biol. Chem.* **1976**, *40*, 303–315. [CrossRef]
8. Bolt, A.J.N.; Purkis, S.W.; Sadd, J.S. A damascone derivative from *Nicotiana tabacum*. *Phytochemistry* **1983**, *22*, 613–614. [CrossRef]
9. Sato, S.; Sasakura, S.; Kobayashi, A.; Nakatani, Y.; Yamanishi, T. Flavor of black tea. Part VI. Intermediate and high boiling components of the neutral fraction. *Agric. Biol. Chem.* **1970**, *34*, 1355–1367. [CrossRef]
10. Beekwilder, J.; van Rossum, H.M.; Koopman, F.; Sonntag, F.; Buchhaupt, M.; Schrader, J.; Hall, R.D.; Bosch, D.; Pronk, J.T.; van Maris, A.J.A.; et al. Polycistronic expression of a β-carotene biosynthetic pathway in *Saccharomyces cerevisiae* coupled to β-ionone production. *J. Biotechnol.* **2014**, *192*, 383–392. [CrossRef] [PubMed]
11. Serra, S.; Fuganti, C.; Brenna, E. Biocatalytic preparation of natural flavours and fragrances. *Trends Biotechnol.* **2005**, *23*, 193–198. [CrossRef] [PubMed]
12. Prelog, V.; Meier, H.L. Untersuchungen über organextrakte und harn. 18. Mitteilung. Über die biochemische oxydation von β-jonon im tierkörper. *Helv. Chim. Acta* **1950**, *33*, 1276–1284. [CrossRef]
13. Mikami, Y.; Watanabe, E.; Fukunaga, Y.; Kisaki, T. Formation of 2S-hydroxy-β-ionone and 4-hydroxy-β-ionone by microbial hydroxylation of β-ionone. *Agric. Biol. Chem.* **1978**, *42*, 1075–1077. [CrossRef]
14. Mikami, Y.; Fukunaga, Y.; Arita, M.; Kisaki, T. Microbial transformation of β-ionone and β-methylionone. *Appl. Environ. Microbiol.* **1981**, *41*, 610–617. [PubMed]
15. Krasnobajew, V.; Helmlinger, D. Fermentation of fragrances: Biotransformation of β-ionone by *Lasiodiplodia theobromae*. *Helv. Chim. Acta* **1982**, *65*, 1590–1601. [CrossRef]
16. Hartman, D.A.; Pontones, M.E.; Kloss, V.F.; Curley, R.W.; Robertson, L.W. Models of retinoid metabolism: Microbial biotransformation of α-ionone and β-ionone. *J. Nat. Prod.* **1988**, *51*, 947–953. [CrossRef] [PubMed]
17. Schoch, E.; Benda, I.; Schreier, P. Bioconversion of α-damascone by *Botrytis cinerea*. *Appl. Environ. Microbiol.* **1991**, *57*, 15–18.
18. Kakeya, H.; Sugai, T.; Ohta, H. Biochemical preparation of optically active 4-hydroxy-β-ionone and its transformation to (S)-6-hydroxy-α-ionone. *Agric. Biol. Chem.* **1991**, *55*, 1873–1876. [CrossRef]
19. Weidmann, V.; Kliewer, S.; Sick, M.; Bycinskij, S.; Kleczka, M.; Rehbein, J.; Griesbeck, A.G.; Zorn, H.; Maison, W. Studies towards the synthetic applicability of biocatalytic allylic oxidations with the lyophilisate of *Pleurotus sapidus*. *J. Mol. Catal. B Enzym.* **2015**, *121*, 15–21. [CrossRef]
20. Gliszczyńska, A.; Gładkowski, W.; Dancewicz, K.; Gabryś, B.; Szczepanik, M. Transformation of β-damascone to (+)-(S)-4-hydroxy-β-damascone by fungal strains and its evaluation as a potential insecticide against aphids *Myzus persicae* and lesser mealworm *Alphitobius diaperinus* Panzer. *Catal. Commun.* **2016**, *80*, 39–43. [CrossRef]
21. Lutz-Wahl, S.; Fischer, P.; Schmidt-Dannert, C.; Wohlleben, W.; Hauer, B.; Schmid, R.D. Stereo- and regioselective hydroxylation of α-ionone by *Streptomyces* strains. *Appl. Environ. Microbiol.* **1998**, *64*, 3878–3881. [PubMed]

22. Maurer, S.C.; Schulze, H.; Schmid, R.D.; Urlacher, V. Immobilisation of P450 BM-3 and an NADP+ cofactor recycling system: Towards a technical application of heme-containing monooxygenases in fine chemical synthesis. *Adv. Synth. Catal.* **2003**, *345*, 802–810. [CrossRef]
23. Litzenburger, M.; Bernhardt, R. Selective oxidation of carotenoid-derived aroma compounds by CYP260B1 and CYP267B1 from *Sorangium cellulosum* So ce56. *Appl. Microbiol. Biotechnol.* **2016**, *100*, 4447–4457. [CrossRef] [PubMed]
24. Venkataraman, H.; Beer, S.B.A.d.; Geerke, D.P.; Vermeulen, N.P.E.; Commandeur, J.N.M. Regio- and stereoselective hydroxylation of optically active α-ionone enantiomers by engineered cytochrome P450 BM3 mutants. *Adv. Synth. Catal.* **2012**, *354*, 2172–2184. [CrossRef]
25. Fuganti, C.; Serra, S.; Zenoni, A. Synthesis and olfactory evaluation of (+)- and (−)-gamma-ionone. *Helv. Chim. Acta* **2000**, *83*, 2761–2768. [CrossRef]
26. Serra, S.; Fuganti, C.; Brenna, E. Synthesis, olfactory evaluation, and determination of the absolute configuration of the 3,4-didehydroionone stereoisomers. *Helv. Chim. Acta* **2006**, *89*, 1110–1122. [CrossRef]
27. Serra, S.; Fuganti, C.; Brenna, E. Two easy photochemical methods for the conversion of commercial ionone alpha into regioisomerically enriched gamma-ionone and gamma-dihydroionone. *Flavour Fragr. J.* **2007**, *22*, 505–511. [CrossRef]
28. Serra, S. An expedient preparation of enantio-enriched ambergris odorants starting from commercial ionone alpha. *Flavour Fragr. J.* **2013**, *28*, 46–52. [CrossRef]
29. Brenna, E.; Fuganti, C.; Serra, S.; Kraft, P. Optically active ionones and derivatives: Preparation and olfactory properties. *Eur. J. Org. Chem.* **2002**, 967–978. [CrossRef]
30. Serra, S.; Fuganti, C. Synthesis of the enantiomeric forms of alpha- and gamma-damascone starting from commercial racemic alpha-ionone. *Tetrahedron Asymmetry* **2006**, *17*, 1573–1580. [CrossRef]
31. Brenna, E.; Fuganti, C.; Serra, S. Synthesis and olfactory evaluation of the enantiomerically enriched forms of 7,11-epoxymegastigma-5(6)-en-9-one and 7,11-epoxymegastigma-5(6)-en-9-ols isomers, identified in passiflora edulis. *Tetrahedron Asymmetry* **2005**, *16*, 1699–1704. [CrossRef]
32. Serra, S.; Barakat, A.; Fuganti, C. Chemoenzymatic resolution of *cis*- and *trans*-3,6-dihydroxy-alpha-ionone. Synthesis of the enantiomeric forms of dehydrovomifoliol and 8,9-dehydrotheaspirone. *Tetrahedron Asymmetry* **2007**, *18*, 2573–2580. [CrossRef]
33. Mazur, M.; Grudniewska, A.; Wawrzeńczyk, C. Microbial transformations of halolactones with *p*-menthane system. *J. Biosci. Bioeng.* **2015**, *119*, 72–76. [CrossRef] [PubMed]
34. Simeo, Y.; Sinisterra, J.V. Biotransformation of terpenoids: A green alternative for producing molecules with pharmacological activity. *Mini-Rev. Org. Chem.* **2009**, *6*, 128–134. [CrossRef]
35. Bhatti, H.N.; Khera, R.A. Biological transformations of steroidal compounds: A review. *Steroids* **2012**, *77*, 1267–1290. [CrossRef]
36. Carballeira, J.D.; Álvarez, E.; Sinisterra, J.V. Biotransformation of cyclohexanone using immobilized *Geotrichum candidum* NCYC49: Factors affecting the selectivity of the process. *J. Mol. Catal. B Enzym.* **2004**, *28*, 25–32. [CrossRef]
37. Bankar, A.V.; Kumar, A.R.; Zinjarde, S.S. Environmental and industrial applications of *Yarrowia lipolytica*. *Appl. Microbiol. Biotechnol.* **2009**, *84*, 847. [CrossRef]
38. Johnson, E.A. *Phaffia rhodozyma*: Colorful odyssey. *Int. Microbiol.* **2003**, *6*, 169–174. [CrossRef]
39. Knapp, H.; Weigand, C.; Gloser, J.; Winterhalter, P. 2-hydroxy-2,6,10,10-tetramethyl-1-oxaspiro[4.5]dec-6-en-8-one: Precursor of 8,9-dehydrotheaspirone in white-fleshed nectarines. *J. Agric. Food Chem.* **1997**, *45*, 1309–1313. [CrossRef]
40. Näf, R.; Velluz, A.; Decorzant, R.; Näf, F. Structure and synthesis of two novel ionone-type compounds identified in quince brandy (*Cydonia oblonga* Mil.). *Tetrahedron Lett.* **1991**, *32*, 753–756. [CrossRef]
41. Serra, S. MnO$_2$/TBHP: A versatile and user-friendly combination of reagents for the oxidation of allylic and benzylic methylene functional groups. *Eur. J. Org. Chem.* **2015**, *2015*, 6472–6478. [CrossRef]
42. Serra, S.; Lissoni, V. First enantioselective synthesis of marine diterpene ambliol-A. *Eur. J. Org. Chem.* **2015**, *2015*, 2226–2234. [CrossRef]
43. Dess, D.B.; Martin, J.C. Readily accessible 12-I-5 oxidant for the conversion of primary and secondary alcohols to aldehydes and ketones. *J. Org. Chem.* **1983**, *48*, 4155–4156. [CrossRef]
44. Audran, G.; Galano, J.M.; Monti, H. Enantioselective synthesis and determination of the absolute configuration of natural (−)-elegansidiol. *Eur. J. Org. Chem.* **2001**, 2293–2296. [CrossRef]

45. Serra, S.; Gatti, F.G.; Fuganti, C. Lipase-mediated resolution of the hydroxy-cyclogeraniol isomers: Application to the synthesis of the enantiomers of karahana lactone, karahana ether, crocusatin C and gamma-cyclogeraniol. *Tetrahedron Asymmetry* **2009**, *20*, 1319–1329. [CrossRef]
46. Kaiser, R.; Lamparsky, D. Inhaltsstoffe des *Osmanthus*-absolues. 1. Mitteilung: 2,5-epoxy-megastigma-6,8-dien. *Helv. Chim. Acta* **1978**, *61*, 373–382. [CrossRef]
47. Tu, V.A.; Kaga, A.; Gericke, K.-H.; Watanabe, N.; Narumi, T.; Toda, M.; Brueckner, B.; Baldermann, S.; Mase, N. Synthesis and characterization of quantum dot nanoparticles bound to the plant volatile precursor of hydroxy-apo-10′-carotenal. *J. Org. Chem.* **2014**, *79*, 6808–6815. [CrossRef]
48. Khachik, F.; Chang, A.-N. Synthesis of (3*S*)- and (3*R*)-3-hydroxy-β-ionone and their transformation into (3*S*)- and (3*R*)-β-cryptoxanthin. *Synthesis* **2011**, *2011*, 509–516. [CrossRef]
49. Takei, Y.; Mori, K.; Matsui, M. Synthesis of a stereoisomeric mixture of 3-hydroxy-α-damascone. *Agric. Biol. Chem.* **1973**, *37*, 2927–2928. [CrossRef]
50. Pascual, A.; Bischofberger, N.; Frei, B.; Jeger, O. Photochemical reactions. 149th communication. Photochemistry of 7,8-dihydro-4-hydroxy-β-ionone and derivatives. *Helv. Chim. Acta* **1988**, *71*, 374–388. [CrossRef]
51. Buschor, D.J.; Eugster, C.H. Synthese der (3*S*,4*R*,3′*S*,4′*R*)- und (3*S*,4*R*,3′*S*,4′*S*)crustaxanthine sowie weiterer verbindungen mit 3,4-dihydroxy-β-endgruppen. *Helv. Chim. Acta* **1990**, *73*, 1002–1021. [CrossRef]
52. Irie, H.; Matsumoto, R.; Nishimura, M.; Zhang, Y. Synthesis of (±)-heritol, a sesquiterpene lactone belonging to the aromatic cadinane group. *Chem. Pharm. Bull.* **1990**, *38*, 1852–1856. [CrossRef]
53. Cooper, R.D.G.; Davis, J.B.; Leftwick, A.P.; Price, C.; Weedon, B.C.L. Carotenoids and related compounds. Part XXXII. Synthesis of astaxanthin, phoenicoxanthin, hydroxyechinenone, and the corresponding diosphenols. *J. Chem. Soc. Perkin Trans. 1* **1975**, 2195–2204. [CrossRef]
54. Serra, S.; Piccioni, O. A new chemo-enzymatic approach to the stereoselective synthesis of the flavors tetrahydroactinidiolide and dihydroactinidiolide. *Tetrahedron Asymmetry* **2015**, *26*, 584–592. [CrossRef]
55. Oritani, T.; Yamamoto, H.; Yamashita, K. Synthesis of (±)-4′-hydroxy-γ-ionylideneacetic acids, fungal biosynthetic intermediates of abscisic acid. *Agric. Biol. Chem.* **1990**, *54*, 125–130. [CrossRef]

Sample Availability: Samples of the compounds **23a**, **25**, **28**, **35** and **36** are available from the authors.

© 2018 by the authors. Licensee MDPI, Basel, Switzerland. This article is an open access article distributed under the terms and conditions of the Creative Commons Attribution (CC BY) license (http://creativecommons.org/licenses/by/4.0/).

Article

Characterization of a Carbonyl Reductase from *Rhodococcus erythropolis* WZ010 and Its Variant Y54F for Asymmetric Synthesis of (S)-N-Boc-3-Hydroxypiperidine

Xiangxian Ying [1,*], Jie Zhang [1], Can Wang [1], Meijuan Huang [1], Yuting Ji [1], Feng Cheng [1], Meilan Yu [2], Zhao Wang [1] and Meirong Ying [3,*]

[1] Key Laboratory of Bioorganic Synthesis of Zhejiang Province, College of Biotechnology and Bioengineering, Zhejiang University of Technology, Hangzhou 310014, China; m15958047548@163.com (J.Z.); m17816035735@163.com (C.W.); meyroline.huang@gmail.com (M.H.); LJ15957189939@163.com (Y.J.); fengcheng@zjut.edu.cn (F.C.); hzwangzhao@163.com (Z.W.)
[2] College of Life Sciences, Zhejiang Sci-Tech Univeristy, Hangzhou 310018, China; meilanyu@zstu.edu.cn
[3] Grain and Oil Products Quality Inspection Center of Zhejiang Province, Hangzhou 310012, China
* Correspondence: yingxx@zjut.edu.cn (X.Y.); hz85672100@163.com (M.Y.); Tel.: +86-571-88320781 (X.Y.)

Academic Editor: Stefano Serra
Received: 11 November 2018; Accepted: 27 November 2018; Published: 28 November 2018

Abstract: The recombinant carbonyl reductase from *Rhodococcus erythropolis* WZ010 (ReCR) demonstrated strict (S)-stereoselectivity and catalyzed the irreversible reduction of N-Boc-3-piperidone (NBPO) to (S)-N-Boc-3-hydroxypiperidine [(S)-NBHP], a key chiral intermediate in the synthesis of ibrutinib. The NAD(H)-specific enzyme was active within broad ranges of pH and temperature and had remarkable activity in the presence of higher concentration of organic solvents. The amino acid residue at position 54 was critical for the activity and the substitution of Tyr54 to Phe significantly enhanced the catalytic efficiency of ReCR. The k_{cat}/K_m values of ReCR Y54F for NBPO, (R/S)-2-octanol, and 2-propanol were 49.17 s^{-1} mM^{-1}, 56.56 s^{-1} mM^{-1}, and 20.69 s^{-1} mM^{-1}, respectively. In addition, the (S)-NBHP yield was as high as 95.92% when whole cells of *E. coli* overexpressing ReCR variant Y54F catalyzed the asymmetric reduction of 1.5 M NBPO for 12 h in the aqueous/(R/S)-2-octanol biphasic system, demonstrating the great potential of ReCR variant Y54F for practical applications.

Keywords: (S)-N-Boc-3-hydroxypiperidine; carbonyl reductase; asymmetric reduction; rational design; *Rhodococcus erythropolis*

1. Introduction

Many natural products and active pharmaceutical ingredients share a common piperidine core, and the introduction of a chiral hydroxyl group on the C3-position of the piperidine ring may alter the bioactivity of the molecule [1–3]. (S)-N-Boc-3-hydroxypiperidine ((S)-NBHP) is a key chiral intermediate in the synthesis of ibrutinib as the inhibitor of Bruton's tyrosine kinase [4]. In the chemical synthesis of (S)-NBHP, employed strategies include the synthesis of racemic 3-hydroxypiperidine followed by chiral resolution and the enantiospecific synthesis of (S)-NBHP from chiral precursors. The former only achieves a maximum yield of 50%, making the process economically unviable, while the latter appears to be limited because of the lengthy procedure, rather poor yields of the products, and the use of potentially hazardous reagents [1,5,6]. Alternatively, the carbonyl-reductase-catalyzed asymmetric reduction of N-Boc-3-piperidone (NBPO) has gained increasing focus due to its mild reaction conditions, high yield, and remarkable enantioselectivity [4,7–9].

Coenzymes are required in carbonyl reductase-catalyzed reactions, and well-established approaches for coenzyme regeneration include the use of a second enzyme and a second substrate (i.e., glucose dehydrogenase and glucose), and the use of the second substrate catalyzed by the same enzyme (i.e., 2-propanol) [10]. Recently, an NADPH-dependent carbonyl reductase from *Saccharomyces cerevisiae* (YDR541C) was employed for the efficient synthesis of (S)-NBHP from NBPO by adopting a biphasic system to alleviate product inhibition and using glucose/glucose dehydrogenase to achieve coenzyme regeneration [8]. The glucose/glucose dehydrogenase system yields to the continuous production of gluconic acid; thus, pH adjustment is needed during the reaction, eventually making the process more complex and forming a large quantity of solid waste salt. Alternatively, the 2-propanol oxidation catalyzed by the same carbonyl reductase was widely used for coenzyme regeneration in order to simplify the operating process and increase the solubility of the substrates [11]. An efficient process catalyzed by the commercially-available ketoreductase KR-110 has been demonstrated to reduce 0.5 M NBPO to render the (S)-NBHP yield of 97.6% after a 24-h reaction [4]. The enzyme KR-110 was heat-sensitive and the substrate inhibition was obviously observed at a substrate concentration of 0.5 M. In addition, the 2-propanol concentration is usually required in excess to increase the product yield. Thus, high concentrations of the co-substrate together with the substrate further aggravate the inhibition of the enzyme activity in the 2-propanol-coupled strategy [4,11].

To overcome the inhibition from the high load of substrate/co-substrate, protein engineering is one of the promising approaches expanding the upper limit of the substrate/co-substrate concentration on a larger preparative scale [12,13]. Variants of the phenylacetaldehyde reductase from *Rhodococcus* sp. ST-10 (PAR) have been constructed through directed evolution, fully converting 200 g/L ethyl 4-chloro-3-oxobutanoate into ethyl (S)-4-chloro-3-hydroxybutyrate in the presence of 15% (v/v) 2-propanol [14,15]. Furthermore, attempts with biphasic catalysis in the presence of water-immiscible organic solvents have demonstrated an intriguing potential for overcoming the inhibition from substrate/co-substrate, increasing the solubility of substrates, easy product removal, decreasing the spontaneous hydrolysis of substrate/product, and avoiding unfavorable equilibria [16–19]. In an aqueous/octanol biphasic system, the biosynthesis process of ethyl (R)-4-chloro-3-hydroxybutyrate using a stereoselective carbonyl reductase from *Burkholderia gladioli* was established, in which 1.2 M ethyl 4-chloro-3-oxobutanoate was completely converted to afford ethyl (R)-4-chloro-3-hydroxybutyrate through the substrate fed-batch strategy [20]. In addition, the integration of protein engineering and medium engineering can further improve the effectiveness of asymmetric reduction at a high substrate load [20–22].

Although several processes for the efficient biosynthesis of (S)-NBHP have been developed, the pivot carbonyl reductases as biocatalysts still lack an in-depth characterization. Our previous genome mining enabled the discovery of chiral ketoreductases from *Rhodococcus erythropolis* WZ010 and the exploration of its application in the synthesis of chiral alcohols [23,24]. Here, a strictly (S)-enantioselective carbonyl reductase from *R. erythropolis* WZ010 (ReCR) and its variant Y54F were characterized for the efficient bioreduction of NBPO to (S)-NBHP, providing a basis for process development with an efficient coenzyme regeneration employing (R/S)-2-octanol or 2-propanol as the co-substrate (Scheme 1).

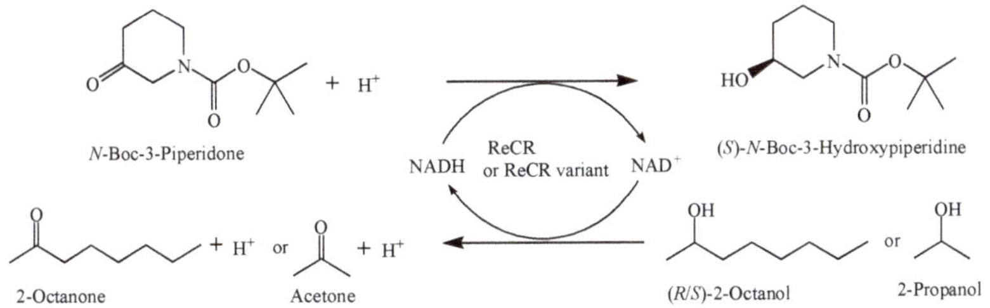

Scheme 1. Asymmetric bioreduction of N-Boc-3-piperidone (NBPO) using (R/S)-2-octanol or 2-propanol as co-substrate for NADH regeneration.

2. Results and Discussion

2.1. Characterization of Recombinant ReCR

The 1044-bp-long gene encoding ReCR was PCR-amplified from the genomic DNA of *R. erythropolis* WZ010 and over-expressed in *E. coli* BL21(DE3) in the form of the recombinant plasmid pEASY-E2-*recr*. The recombinant ReCR with C-terminal His-tag was subsequently purified by Ni-NTA chromatography. The gene *recr* encoded 348 amino acids with a deduced mass of 36.17 kDa, and the purified recombinant ReCR was verified with a single band of around 44 kDa by SDS-PAGE (Figure 1). The encoded amino acid sequence of ReCR displayed a 98% identity to that of PAR or alcohol dehydrogenase from *R. erythropolis* DSM 43297 (ReADH) [25–28], with five amino acids Arg67, Ser94, Lys110, Ser233, and Arg336 in ReCR different from Lys67, Asn94, Gln110, Lys233, and Gly336 in PAR or ReADH (Figure 2). The structure-related sequence alignment revealed that the enzyme belonged to the superfamily of zinc-containing alcohol dehydrogenases and had all conserved residues for the binding of catalytic and structural zinc ions [29]. It should be noted that the activity of the enzyme was severely inhibited by the exogenous zinc ion (Table S1), similar to what was observed with other zinc-containing alcohol dehydrogenases [24,30].

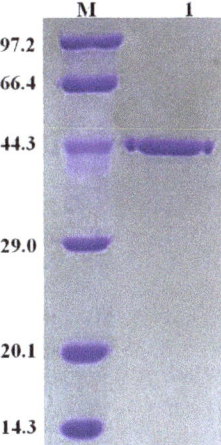

Figure 1. SDS-PAGE (12.5%) analysis of the purified recombinant ReCR. Lane 1, 2 μg purified ReCR with C-terminal His-tag; lane M, molecular weight marker. Coomassie Brilliant Blue R-250 was used to visualize the protein bands in the SDS-PAGE gel.

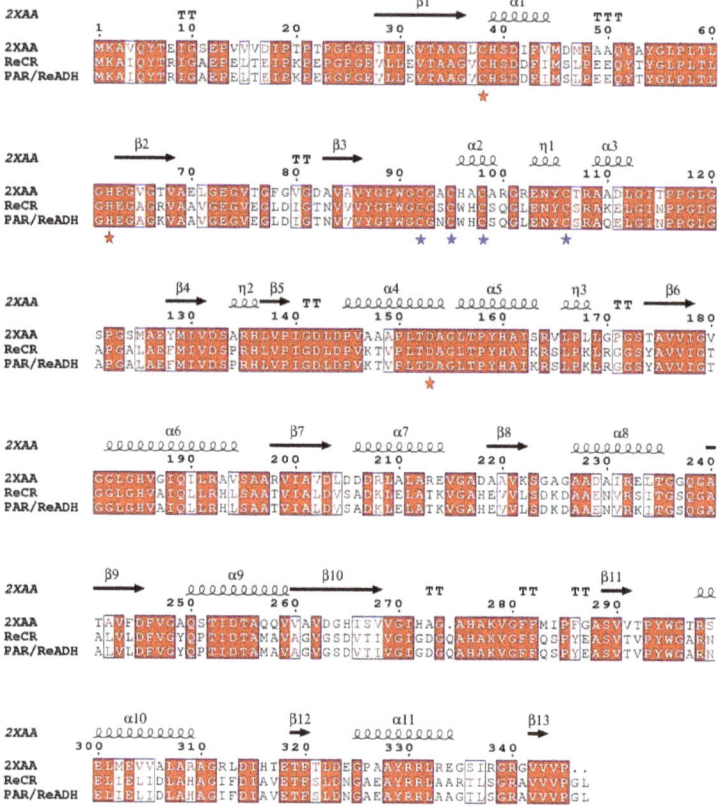

Figure 2. Structure-related sequence alignment between ReCR and its homologous proteins. 2XAA, PDB code of alcohol dehydrogenase from *Rhodococcus ruber* DSM 44541; PAR, alcohol dehydrogenase from *Rhodococcus* sp. ST-10 (GenBank accession No.: AB020760.3); ReADH, alcohol dehydrogenase from *R. erythropolis* DSM 43297 (GenBank accession No.: AY161280.1). The amino acid sequences of both PAR and ReADH are identical. Shown above the alignments are elements of the secondary structure of 2XAA. The numbering shown is from 2XAA. Red stars, putative catalytic residues; blue stars, residues for the coordination of structural zinc. Strictly conserved residues are highlighted with red boxes.

The recombinant ReCR was strictly NAD^+-dependent, since the enzyme activity was not detectable when NADP(H) was used as a coenzyme. The effect of pH on the activity was investigated within the pH range of 5.5–10.5. The maximum activities for NBPO reduction and (R/S)-2-octanol oxidation were observed at pH 6.0 and 10.0, respectively (Figure 3A), indicating that ReCR-catalyzed oxidation/reduction was pH-dependent [24]. The optimal temperature was 60 °C for NBPO reduction and 50 °C for (R/S)-2-octanol oxidation (Figure 3B). The enzyme activity in NBPO reduction was stable at 35 °C, whereas the remaining activity decreased to 50% of the initial activity after heat treatment at 60 °C for 1.5 h or 55 °C for 6.5 h (Figure 4A), demonstrating that its thermostability was superior to the heat-sensitive enzyme KR-110 [4]. Among the tested organic solvents, 20% (v/v) 2-propanol drastically decreased the activity of ReCR, similar to the performance of PAR in the presence of >10% (v/v) 2-propanol [18]. In contrast to 20% (v/v) 2-propanol, the enzyme displayed higher stability after 3.5 h incubation with 40% (v/v) (R/S)-2-octanol (Figure 4B).

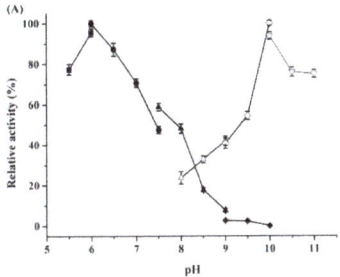
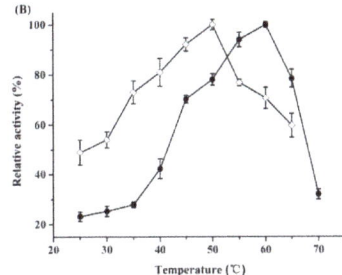

Figure 3. Effect of pH (**A**) and temperature (**B**) on the activity of recombinant ReCR. The relative activities of 100% represent 85.8 U/mg for NBPO reduction (solid symbols) and 88.3 U/mg for (R/S)-2-octanol oxidation (open symbols). The buffers 2-(N-morpholino)ethanesulfonic acid (MES, ■), piperazine-1,4-bisethanesulfonic acid(PIPES, ●), Tris-HCl (▲), and 3-(cyclohexylamino)-2-hydroxy-1-propanesulfonic acid (CAPSO, ◆) were used for the reduction reaction, while the buffers Tris-HCl (△), CAPSO (◇), and 3-(cyclohexylamino)-1-propanesulfonic acid (CAPS, ▽) were used for the oxidation reaction.

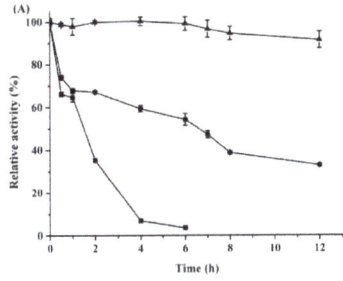
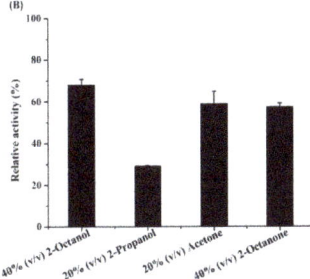

Figure 4. The stability of ReCR against heat (**A**) and organic solvents (**B**). Symbols: (■) for 60 °C, (●) for 55 °C, (▲) for 35 °C. The relative activity of 100% represents 85.8 U/mg for NBPO reduction. The enzyme was incubated with organic solvent (40% (v/v) (R/S)-2-octanol, 40% (v/v) 2-octanone, 20% (v/v) 2-propanol, or 20% (v/v) acetone) at 35 °C for 3.5 h prior to the stability test against organic solvent.

The substrate specificity of ReCR was tested using a set of alcohols and ketones (Table 1). Among the tested substrates, the enzyme exhibited the highest activities with 2,3-butanedione in the ketone reduction and (R/S)-2-octanol in the alcohol oxidation. The purified ReCR presented an activity of 85.8 U/mg towards NBPO reduction at pH 6.0 and 60 °C. Distinct from PAR and its variants [14], the activity of ReCR toward N-Boc-3-pyrrolidone reduction was relatively low. Particularly, the activity towards the oxidation of either (S)- or (R)-NBHP was not detectable at various temperatures (25–75 °C) and pHs (6.0–10.0), suggesting that the ReCR-catalyzed NBPO reduction was irreversible. A similar case was the secondary alcohol dehydrogenase SdcA from R. erythropolis DSM 44534 catalyzing the irreversible (S)-2-octanol oxidation [31]. The K_m and k_{cat}/K_m values for NBPO were 1.74 mM and 35.98 s^{-1} mM^{-1}, respectively (Table 2). The k_{cat}/K_m value for (R/S)-2-octanol and 2-propanol was 13.04 s^{-1} mM^{-1} and 9.74 s^{-1} mM^{-1}, respectively, implying that the use of (R/S)-2-octanol or 2-propanol as a co-substrate could be feasible to regenerate NADH in the NBPO reduction.

Table 1. Substrate spectrum of recombinant ReCR against ketones and alcohols [a].

Substrate	Relative Activity (%)	Substrate	Relative Activity (%)
N-Boc-3-Piperidone	100.0 [b] ± 2.6	(R/S)-2-Octanol	100.0 [c] ± 1.6
2,3-Butanedione	189.0 ± 3.4	(R/S)-2-Pentanol	61.8 ± 2.3
2-Octanone	169.2 ± 2.9	2-Propanol	47.4 ± 0.5
p-Bromoacetophenone	143.9 ± 4.3	(R/S)-2-Butanol	43.8 ± 1.1
Acetoin	47.2 ± 0.7	DL-1-Phenylethanol	31.5 ± 2.1
β-Ionone	34.8 ± 1.2	Cyclohexanol	8.0 ± 1.0
4-Hydroxy-2-butanone	31.8 ± 1.1	2-Buten-1-ol	6.8 ± 0.2
3-Octen-2-one	25.7 ± 0.7	(S)-N-Boc-3-Pyrrolidinol	2.7 ± 0.4
Acetophenone	25.3 ± 1.0	(S)-N-Boc-3-Hydroxypiperidine	0
Hydroxyacetone	23.6 ± 0.6	(R)-N-Boc-3-Hydroxypiperidine	0
N-Boc-3-Pyrrolidone	9.2 ± 0.5		
Acetone	4.8 ± 0.3		
2-Bromoacetophenone	1.8 ± 0.1		

[a] Data present mean values ± SD from two independent experiments. [b] Relative activity of 100% represents 85.8 U/mg for NBPO reduction at pH 6.0 and 60 °C; [c] Relative activity of 100% represents 88.3 U/mg for (R/S)-2-octanol oxidation at pH 10.0 and 50 °C.

Table 2. Kinetic parameters of recombinant ReCR [a].

Substrate	Coenzyme (mM)	V_{max} (U mg^{-1})	K_m (mM)	k_{cat} (s^{-1})	k_{cat}/K_m (s^{-1} mM^{-1})
NBPO	NADH (0.4)	103.57 ± 2.46	1.74 ± 0.08	62.61 ± 1.49	35.98 ± 0.86
(S)-NBHP	NAD$^+$ (0.4)	ND [b]	ND [b]	ND [b]	ND [b]
Acetone	NADH (0.4)	66.30 ± 3.27	46.06 ± 2.62	40.08 ± 1.98	0.87 ± 0.04
2-Propanol	NAD$^+$ (0.4)	23.54 ± 0.27	1.46 ± 0.06	14.22 ± 0.16	9.74 ± 0.11
2-Octanone	NADH (0.4)	235.54 ± 5.95	3.29 ± 0.05	142.38 ± 3.11	43.28 ± 0.95
(R/S)-2-Octanol	NAD$^+$ (0.4)	106.57 ± 2.74	4.94 ± 0.45	64.42 ± 1.66	13.04 ± 0.34

[a] Data present mean values ± SD from three independent experiments. [b] ND, not detectable.

2.2. Rational Design and Characterization of ReCR Variant Y54F

For the in-depth characterization, attempts of rational design of ReCR were conducted to improve its activity. The ReCR homology model was built based on the X-ray crystal structure of ADH-A from *Rhodococcus ruber* (PDB: 2XAA). Sequence identity of ReCR towards ADH-A was 60%. The QMEAN and Z-score values were used for the quality evaluation of the models. The QMEAN and Z-score values of the ReCR homology model were 0.822 and 0.533, respectively, which indicated satisfactory quality. In Ramachandran Plot analysis, 91.5% of residues were located in a favorable region, and only 0.4% were found in the sterically disallowed region. This ReCR homology model was selected for subsequent docking studies.

Furthermore, substrate docking was employed to predict potentially beneficial amino acid positions on ReCR. Figure 5A shows that NBPO was ideally accommodated in the ligand binding pocket of ReCR composed by zinc ion, NADH, and Tyr54 (in the vicinity of the entrance to the active site). Similar to the binding mode of ADH-A with the substrate [29], the carbonyl oxygen atom of NBPO in ReCR was bound to the Zn^{2+} ion with a distance of 4.1 Å, and the carbonyl carbon atom was in close proximity to the C4-atom of NADH. Thus, the hydride was transferred onto the *re*-face of the carbonyl group, consistent with the strict (S)-enantioselectivity of ReCR. On the other hand, the bulky Boc group of NBPO was close to the hydroxyl group of Tyr54 (distance of 4.3 Å between the hydroxyl oxygen of Tyr and the tertiary carbon of the Boc group), which might cause a steric hindrance during the substrate binding (Figure 5). Therefore, Tyr54 was selected to be mutated to Phe.

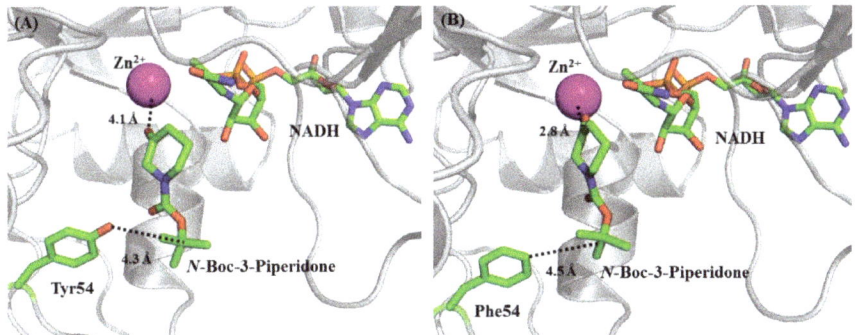

Figure 5. Protein-ligand structures of ReCR with NBPO (**A**) and ReCR Y54F with NBPO (**B**). ReCR and ReCR Y54F are represented in cartoon format. Tyr54, Phe54, NADH, and NBPO are highlighted in sticks. The zinc ion is shown as a magenta sphere.

As anticipated, the substitution of Tyr54 to Phe significantly improved the catalytic performance of ReCR, implying that the amino acid residue at position 54 could be critical for the enzyme activity. In the ketone reduction, the k_{cat}/K_m values of ReCR Y54F for NBPO (49.17 s^{-1} mM^{-1}), acetone (1.47 s^{-1} mM^{-1}), and 2-octanone (53.21 s^{-1} mM^{-1}) were 1.37, 1.69, and 1.23 times higher than those of ReCR (35.98 s^{-1} mM^{-1}, 0.87 s^{-1} mM^{-1}, and 43.28 s^{-1} mM^{-1}), respectively (Tables 2 and 3). In the alcohol oxidation, the k_{cat}/K_m values of ReCR Y54F for (R/S)-2-octanol (56.56 s^{-1} mM^{-1}) and 2-propanol (20.69 s^{-1} mM^{-1}) were 4.34 and 2.12 times higher than those of ReCR (13.04 s^{-1} mM^{-1} and 9.74 s^{-1} mM^{-1}), respectively (Tables 2 and 3). Although the K_m value of ReCR Y54F for NBPO (1.74 mM) was similar to that of ReCR, the K_m values of ReCR Y54F for other tested substrates were lowered to a certain extent.

Table 3. Kinetic parameters of ReCR variant Y54F [a].

Substrate	Coenzyme (mM)	V_{max} (U mg^{-1})	K_m (mM)	k_{cat} (s^{-1})	k_{cat}/K_m (s^{-1} mM^{-1})
NBPO	NADH (0.4)	140.72 ± 6.52	1.73 ± 0.05	85.07 ± 3.94	49.17 ± 2.28
(S)-NBHP	NAD$^+$ (0.4)	ND [b]	ND [b]	ND [b]	ND [b]
Acetone	NADH (0.4)	90.46 ± 1.69	37.32 ± 0.56	54.68 ± 1.02	1.47 ± 0.03
2-Propanol	NAD$^+$ (0.4)	35.29 ± 0.88	1.03 ± 0.05	21.32 ± 0.53	20.69 ± 0.51
2-Octanone	NADH (0.4)	273.75 ± 7.58	3.11 ± 0.12	165.48 ± 4.58	53.21 ± 1.47
(R/S)-2-Octanol	NAD$^+$ (0.4)	128.19 ± 3.12	1.37 ± 0.08	77.49 ± 1.88	56.56 ± 1.37

[a] Data present mean values ± SD from three independent experiments. [b] ND, not detectable.

Consistently with kinetic parameters, the productivity of asymmetric bioreduction of NBPO was significantly enhanced when whole cells overexpressing ReCR Y54F instead of ReCR were used as biocatalyst (Table 4). In contrast to the free enzyme, the use of a whole-cell biocatalyst was chosen because of higher enzyme stability and simpler procedure of biocatalyst preparation [4,11,32]. Both 2-propanol and (R/S)-2-octanol were investigated as co-substrates for the NADH regeneration. In the presence of 10% (v/v) 2-propanol, the bioreduction of 0.5 M NBPO catalyzed by whole cells overexpressing ReCR Y54F gave a (S)-NBHP yield of 98.08% after 12 h, which was 1.34 times higher than that of ReCR (72.15%). The whole-cell biphasic system has been demonstrated to be effective at a higher substrate load, in which (R/S)-2-octanol instead of 2-propanol was used not only as co-substrate for coenzyme regeneration but also as the organic phase for the substrate reservoir and product sink [33,34]. In the aqueous/(R/S)-2-octanol biphasic system, the (S)-NBHP yield was increased from 77.78% to 95.92% when ReCR Y54F replaced ReCR in the whole-cell biocatalyst. The corresponding total turnover number value of 1199, the calculated space-time yield of 579.15 g L^{-1} day^{-1}, and the remarkable stereoselectivity ($e.e._p$ > 99.9%) together with the substrate concentration (up to 1.5 M) demonstrated a great potential of ReCR variant Y54F in the practical synthesis of (S)-NBHP.

Table 4. Asymmetric reduction of N-Boc-3-piperidone catalyzed by whole cells overexpressing ReCR or ReCR Y54F [a].

Enzyme [b]	Substrate (M)	Co-substrate (v/v)	Yield (%)	$e.e._p$ (%) [c]
ReCR	NBPO, 0.5	2-Propanol, 10%	72.15 ± 3.51	>99.9 (S)
ReCR Y54F	NBPO, 0.5	2-Propanol, 10%	98.08 ± 1.65	>99.9 (S)
ReCR	NBPO, 1.5	(R/S)-2-Octanol, 60%	77.78 ± 2.23	>99.9 (S)
ReCR Y54F	NBPO, 1.5	(R/S)-2-Octanol, 60%	95.89 ± 2.37	>99.9 (S)

[a] Data present mean values ± SD from two independent experiments. [b] Whole cells overexpressing ReCR or ReCR variant Y54F. [c] The $e.e._p$ value (>99.9%) means that no (R)-NBHP peak was detected during GC analyses.

3. Materials and Methods

3.1. Strain and Growth Condition

The strain R. erythropolis WZ010 was deposited in the China Center for Type Culture Collection (CCTCC M 2011336) and used as the donor of the gene recr encoding the carbonyl reductase ReCR [35]. The host strains E. coli Trans1-T1 and E. coli BL21(DE3) were used for the purposes of cloning and over-expression, respectively. Both R. erythropolis WZ010 and E. coli strains were cultured at 30 °C and 200 rpm for 24 h in Luria-Bertani (LB) medium with a NaCl concentration of 5 g/L, unless stated otherwise.

3.2. Construction, Expression, and Purification of Recombinant Enzyme ReCR

The gene recr was PCR-amplified from the genomic DNA of R. erythropolis WZ010 using forward and reverse primers: recrF1 (5'-ATGAAGGCAATCCAGTACAC-3') and recrR1 (5'-CTACAGACCAG GGACCACA-3'). The PCR conditions were listed as follows: denaturalization, 94 °C for 4 min; 30 cycles of 94 °C for 30 s, 53.5 °C for 30 s, and 72 °C for 1 min; and the final extension, 72 °C for 10 min. According to TA cloning strategy from the instructions of the pEASY-E2 expression kit (TransGen Biotech Co., Ltd., Beijing, China), the PCR product was subcloned into the expression vector pEASY-E2 to form the recombinant vector pEASY-E2-recr with the C-terminal His-tag. The recombinant plasmid was then transformed into Trans1-T1 competent cells and the recombinant cells were cultured at 37 °C and 200 rpm in LB medium with 100 µg/mL ampicillin (Amp). The recombinant cell named as E. coli Trans1-T1/pEASY-E2-recr was selected by colony PCRs and the recombinant plasmid pEASY-E2-recr was further extracted and verified by DNA sequencing (Sunny Biotechnology, Shanghai, China).

The recombinant plasmid pEASY-E2-recr was extracted and then transformed into E. coli BL21(DE3) competent cells. The positive recombinant cell named as E. coli BL21(DE3)/pEASY-E2-recr was cultured at 37 °C and 200 rpm in LB medium with 100 µg/mL Amp. When the OD_{600} reached 0.6, isopropyl β-D-1-thiogalactopyranoside (IPTG) was added to the culture at a final concentration of 0.3 mM, and the temperature was maintained at 20 °C. After 20 h incubation, the E. coli cells were harvested by centrifugation and the expression level was analyzed by sodium dodecyl sulfate-polyacrylamide gel electrophoresis (SDS-PAGE). Following the same procedure in the study of 2,3-butanediol dehydrogenase from R. erythropolis WZ010 [24], the recombinant ReCR with C-terminal His-tag was purified to homogeneity by nickel affinity chromatography, desalted with 50 mM Tris-HCl (pH 8.0) by ultrafiltration, and stored at −20 °C for further characterization. The subunit molecular mass and purity of ReCR were verified by SDS-PAGE as described previously [36].

3.3. Enzyme Activity Assays and Characterization of Recombinant ReCR

The ReCR enzyme activity was measured by the reduction of NAD^+ or oxidation of NADH at 340 nm (ε_{340} = 6.3 mM^{-1} cm^{-1}). Unless otherwise specified, the standard enzyme activity assay for the ketone reduction was performed at 60 °C in duplicate using the assay mixture (2.5 mL) containing 10 mM NBPO, 0.4 mM NADH, and 50 mM PIPES buffer (pH 6.0). The standard assay mixture (2.5 mL) for the alcohol oxidation at 50 °C contained 50 mM (R/S)-2-octanol, 0.4 mM NAD^+, and 50 mM CAPSO buffer (pH 10.0). Unless stated otherwise, the reduction and oxidation reactions were initiated by the addition of 5 µg purified enzyme, respectively. One unit of activity was defined as the amount of enzyme that oxidized or reduced 1 µmol NADH or NAD^+ per minute under optimal pH and temperature. The protein concentrations of ReCR samples were determined using the Bradford reagent with bovine serum albumin as the standard protein.

The optimal temperature of ReCR activity was determined at a series of temperatures ranging from 25 to 70 °C using 50 mM PIPES buffer (pH 6.0) for NBPO reduction or 50 mM CAPSO buffer (pH 10.0) for (R/S)-2-octanol oxidation. The optimal pH of ReCR activity was determined over a range of pH from 5.5 to 11.0 at 60 °C for NBPO reduction or 50 °C for (R/S)-2-octanol oxidation. The buffers (50 mM) used were 2-(N-morpholino)ethanesulfonic acid (MES, pH 5.5–6.0), piperazine-1,4-bisethanesulfonic acid (PIPES, pH 6.1–7.5), Tris-HCl (pH 7.5–9.0), 3-(cyclohexylamino)-2-hydroxy-1-propanesulfonic acid (CAPSO, pH 9.0–10.0), and 3-(cyclohexylamino)-1-propanesulfonic acid (CAPS, pH 10.0–11.0). All the pH values of the buffers used were determined at 25 °C using a Mettler Toledo FE20 FiveEasy pH Meter (Mettler-Toledo (Schweiz) GmbH, Greifensee, Switzerland).

The thermostability of the ReCR was investigated by determining its residual activities when the enzyme samples were incubated at 35 °C, 55 °C, or 60 °C. To determine the stability in the presence of organic solvents, the enzyme was incubated with organic solvent at 35 °C for 3.5 h and then the residual activities were assayed for NBPO reduction. The determination of kinetic constants for ReCR was carried out using different substrates. The substrates were NBPO (0–20 mM), acetone (0–1 M), 2-propanol (0–70 mM), 2-octanone (0–30 mM), and (R/S)-2-octanol (0–20 mM). Apparent values of K_m and V_{max} were calculated using a non-linear regression curve fitting to the Michaelis-Menten equation with the software Origin 8.0 (OriginLab Corporation, Northampton, UK). Data of kinetic parameters present mean values ± SD from three independent experiments.

3.4. Asymmetric Reduction of NBPO Catalyzed by Whole Cells of E. coli BL21(DE3)/pEASY-E2-recr

The asymmetric reduction of NBPO was carried out using (R/S)-2-octanol or 2-propanol as a co-substrate for the coenzyme regeneration. In the case of 2-propanol as the co-substrate, the reaction mixture (5 mL) contained 0.5 M NBPO, 10% (v/v) 2-propanol, 0.4 mM NAD^+, and 0.4 g wet cells in 50 mM Tris-HCl buffer (pH 8.0). In the aqueous/(R/S)-2-octanol biphasic system, the reaction mixture (5 mL) contained 1.5 M NBPO, 60% (v/v) (R/S)-2-octanol, 1.2 mM NAD^+, and 1.2 g wet cells in 50 mM Tris-HCl buffer (pH 8.0). The reactions were carried out in a C76 Water Bath Shaker (New Brunswick, Edison, NJ, USA) at 35 °C and 300 rpm for 12 h.

The reaction mixture was extracted with 5 mL of ethyl acetate under strong vibration. The organic phase in the samples was separated by centrifugation and dehydrated with anhydrous sodium sulfate; then, 1 µL dehydrated sample was applied onto the injector (250 °C) for GC analysis. The reactants were determined with an Agilent 6890N (Santa Clara, CA, USA) gas chromatograph equipped with a chiral GC column (BGB174, 30 m × 250 µm × 0.25 µm). The temperature program for GC analysis was set as follows: 5 °C/min from 100 °C to 125 °C, hold 3 min; 2 °C/min to 140 °C, hold 8 min; 1 °C/min to 150 °C. The peak areas were quantitated using specific external standards. The standards NBPO, (S)-NBHP, and (R)-NBHP were purchased from Sigma-Aldrich Corporation (Shanghai, China). Retention times of the reactants were listed as follows: 26.997 min for NBPO, 28.452 min for (S)-NBHP, and 28.739 min for (R)-NBHP (Figure S1). Specifically, the (S)-NBHP peak was further determined by GC-MS analysis (Figure S2).

3.5. Construction, Characterization, and Docking Analysis of ReCR Variant Y54F

Site-specific mutagenesis was carried out by inverse PCR using native pEASY E2-recr as a template and a pair of primers Y54F F1 (5'-TACACCTTCGGCCTTCCTCTCACGC-3') and Y54F

R1 (5′-AAGGCCGAAGGTGTACTGCTCCTCG-3′) under conditions as follows: denaturation, 95 °C for 2 min; 30 cycles of 95 °C for 20 s, 68 °C for 20 s, and 72 °C for 3 min; and the final extension, 72 °C for 8 min. The PCR product was digested at 37 °C for 2 h to digest the native template with the help of *Dpn* I. The digested product was directly transformed into *E. coli* BL21(DE3) competent cells. The positive recombinant cells were cultured at 37 °C and 200 rpm in LB medium with 100 µg/mL Amp. The recombinant cell named as *E. coli* Trans1-T1/pEASY-E2-*recr-mut* was selected by colony PCRs and the recombinant plasmid pEASY-E2-*recr-mut* was further extracted and verified by DNA sequencing (Sunny Biotechnology, Shanghai, China). Following the same procedure for the recombinant ReCR, the positive recombinant cell named as *E. coli* BL21(DE3)/pEASY-E2-*recr-mut* was obtained and the ReCR variant Y54F was purified for further characterization including kinetic parameters and catalytic performance in NBPO reduction.

The homology model of ReCR was built on the X-ray crystallographic structures of ADH-A from *Rhodococcus ruber* (PDB: 2XAA, resolution of 2.8 Å) by HHpred server [37]. Water molecules, ligands, and other hetero atoms (except the NAD^+ coenzyme and the zinc ion) were removed from the protein molecule. The coenzyme was remodeled as NADH. The charge of the catalytic zinc ion was assigned to +2, and the ligating side chain of Cys 38 was set as deprotonated and negatively charged. For the homology model of ReCR Y54F, the substitution of Tyr54 to Phe was introduced by FoldX [38]. A structure energy minimization of the proteins was performed to remove improper torsions of the side-chain conformation and correct the covalent geometry. The ligand molecule structures (NBPO and NADH) were directly drawn in ChemBioDraw and followed by an energy minimization. Global docking was performed using AutoDock Vina under the default docking parameters [39]. Point charges were initially assigned according to the AMBER03 force field [40], and then damped to mimic the less polar Gasteiger charges. Subsequently, local docking was executed to predict the binding energy and fine-tune the ligand placement in the binding site.

3.6. Nucleotide Sequence Accession Number

The nucleotide sequence of ReCR has been submitted to the GenBank database under the accession number of KX827723.

4. Conclusions

The enzyme ReCR showed high specific activity, moderate thermostability, and strict (*S*)-stereoselectivity for asymmetric bioreduction of NBPO to (*S*)-NBHP. The NAD(H)-specific enzyme was active over broad pH and temperature ranges, and tolerated a higher concentration of organic solvents, offering greater flexibility in practical biocatalysis. Particularly, the reduction of NBPO to (*S*)-NBHP was irreversible, which was kinetically in favor of both coenzyme regeneration and formation of (*S*)-NBHP. The substitution of Tyr54 to Phe further improved the catalytic efficiency of ReCR including kinetic parameters and the productivity of (*S*)-NBHP. The k_{cat}/K_m values of ReCR Y54F for NBPO (49.17 s^{-1} mM^{-1}), (*R*/*S*)-2-octanol (56.56 s^{-1} mM^{-1}), and 2-propanol (20.69 s^{-1} mM^{-1}) were 1.37, 4.34, and 2.12 times higher than those of ReCR (35.98 s^{-1} mM^{-1}, 13.04 s^{-1} mM^{-1}, and 9.74 s^{-1} mM^{-1}), respectively. Furthermore, the (*S*)-NBHP yield was increased from 77.78% to 95.89% in the aqueous/(*R*/*S*)-2-octanol biphasic system when asymmetric reduction of 1.5 M NBPO was catalyzed for 12 h by whole cells of *E. coli* overexpressing ReCR Y54F instead of ReCR. Taken as a whole, ReCR variant Y54F has a great potential in the asymmetric synthesis of (*S*)-NBHP using (*R*/*S*)-2-octanol or 2-propanol as a co-substrate.

Supplementary Materials: The following are available online at http://www.mdpi.com/1420-3049/23/12/3117/s1, Figure S1: Gas chromatograph analysis for standards *N*-Boc-3-piperidone (A), (*S*)-*N*-Boc-3-hydroxypiperidine and (*R*)-*N*-Boc-3-hydroxypiperidine; Figure S2: Gas chromatograph-mass spectrometry analysis of (*S*)-*N*-Boc-3-hydroxypiperidine in asymmetric reduction of *N*-Boc-3-piperidone; Figure S3: The Michaelis-Menten kinetics of ReCR; Figure S4. The Michaelis-Menten kinetics of ReCR variant Y54F; Table S1: Effect of metal ions, EDTA, dithiothreitol and sodium iodoacetate on the activity of recombinant ReCR.

Author Contributions: Conceptualization, X.Y. and M.Y. (Meirong Ying); Data curation, X.Y. and M.Y. (Meilan Yu); Formal analysis, X.Y., F.C. and M.Y. (Meirong Ying); Funding acquisition, X.Y. and M.Y. (Meilan Yu); Investigation, J.Z., C.W., M.H., Y.J. and F.C.; Supervision, Z.W.; Writing – original draft, X.Y. and M.Y. (Meirong Ying).

Funding: This work was supported by the Natural Science Fundation of Zhejiang Province, China (No. LY17B020012) and the National Natural Science Fundation of China (No. 21405140).

Conflicts of Interest: The authors declare no conflict of interest.

References

1. Babu, M.S.; Raghunadh, A.; Ramulu, K.; Dahanukar, V.H.; Kumar, U.K.S.; Dubey, P.K. A practical and enantiospecific synthesis of (−)-(R)- and (+)-(S)-piperidin-3-ols. *Helv. Chim. Acta* **2014**, *97*, 1507–1515. [CrossRef]
2. Vitaku, E.; Smith, D.T.; Njardarson, J.T. Analysis of the structural diversity, substitution patterns, and frequency of nitrogen heterocycles among U.S. FDA approved pharmaceuticals. *J. Med. Chem.* **2014**, *57*, 10257–10274. [CrossRef] [PubMed]
3. Chen, L.-F.; Zhang, Y.-P.; Fan, H.-Y.; Wu, K.; Lin, J.-P.; Wang, H.-L.; Wei, D.-Z. Efficient bioreductive production of (R)-N-Boc-3-hydroxypiperidine by a carbonyl reductase. *Catal. Commun.* **2017**, *97*, 5–9. [CrossRef]
4. Ju, X.; Tang, Y.; Liang, X.; Hou, M.; Wan, Z.; Tao, J. Development of a biocatalytic process to prepare (S)-N-Boc-3-hydroxypiperidine. *Org. Process Res. Dev.* **2014**, *18*, 827–830. [CrossRef]
5. Amat, M.; Llor, N.; Huguet, M.; Molins, E.; Espinosa, E.; Bosch, J. Unprecedented oxidation of a phenylglycinol-derived 2-pyridone: Enantioselective synthesis of polyhydroxypiperidines. *Org. Lett.* **2001**, *3*, 3257–3260. [CrossRef] [PubMed]
6. Zhang, Y.-J.; Zhang, W.-X.; Zheng, G.-W.; Xu, J.-H. Identification of an ε-keto ester reductase for the efficient synthesis of an (R)-α-lipoic acid precursor. *Adv. Synth. Catal.* **2015**, *357*, 1697–1702. [CrossRef]
7. Lachereez, R.; Pardo, D.G.; Cossy, J. *Daucus carota* mediated-reduction of cyclic 3-oxo-amines. *Org. Lett.* **2009**, *11*, 1245–1248. [CrossRef] [PubMed]
8. Chen, L.-F.; Fan, H.-Y.; Zhang, Y.-P.; Wu, K.; Wang, H.-L.; Lin, J.-P.; Wei, D.-Z. Development of a practical biocatalytic process for (S)-N-Boc-3-hydroxypiperidine synthesis. *Tetrahedron Lett.* **2017**, *58*, 1644–1650. [CrossRef]
9. Xu, G.-P.; Wang, H.-B.; Wu, Z.-L. Efficient bioreductive production of (S)-N-Boc-3-hydroxypiperidine using ketoreductase ChKRED03. *Proc. Biochem.* **2016**, *51*, 881–885. [CrossRef]
10. Hummel, W.; Groger, H. Strategies for regeneration of nicotinamide coenzymes emphasizing self-sufficient closed-loop recycling systems. *J. Biotechnol.* **2014**, *191*, 22–31. [CrossRef] [PubMed]
11. Stamper, W.; Kosjek, B.; Faber, K.; Kroutil, W. Biocatalytic asymmetric hydrogen transfer employing *Rhodococcus ruber* DSM 44541. *J. Org. Chem.* **2003**, *68*, 402–406. [CrossRef] [PubMed]
12. Huang, L.; Ma, H.-M.; Yu, H.-L.; Xu, J.-H. Altering the substrate specificity of reductase CgKR1 from *Candida glabrata* by protein engineering for bioreduction of aromatic α-keto esters. *Adv. Synth. Catal.* **2014**, *356*, 1943–1948. [CrossRef]
13. Turner, N.J.; O'Reilly, E. Biocatalytic retrosynthesis. *Nat. Chem. Biol.* **2013**, *9*, 285–288. [CrossRef] [PubMed]
14. Itoh, N.; Isotani, K.; Nakamura, M.; Inoue, K.; Isogai, Y.; Makino, Y. Efficient synthesis of optically pure alcohols by asymmetric hydrogen-transfer biocatalysis: Application of engineered enzymes in a 2-propanol-water medium. *Appl. Microbiol. Biotechnol.* **2012**, *93*, 1075–1085. [CrossRef] [PubMed]
15. Makino, Y.; Inoue, K.; Dairi, T.; Itoh, N. Engineering of phenylacetaldehyde reductase for efficient substrate conversion in concentrated 2-propanol. *Appl. Environ. Microbiol.* **2005**, *71*, 4713–4720. [CrossRef] [PubMed]
16. Au, S.K.; Bommarius, B.R.; Bommarius, A.S. Biphasic reaction system allows for conversion of hydrophobic substrates by amine dehydrogenases. *ACS Catal.* **2014**, *4*, 4021–4026. [CrossRef]
17. De Gonzalo, G.; Lavandera, I.; Faber, K.; Kroutil, W. Enzymatic reduction of ketones in "micro-aqueous" media catalyzed by ADH-A from *Rhodococcus ruber*. *Org. Lett.* **2007**, *9*, 2163–2166. [CrossRef] [PubMed]
18. Itoh, N.; Matsuda, M.; Mabuchi, M.; Dairi, T.; Wang, J. Chiral alcohol production by NADH-dependent phenylacetaldehyde reductase coupled with in situ regeneration of NADH. *Eur. J. Biochem.* **2002**, *269*, 2394–2402. [CrossRef] [PubMed]
19. Xu, G.-C.; Tang, M.-H.; Ni, Y. Asymmetric synthesis of lipitor chiral intermediate using a robust carbonyl reductase at high substrate to catalyst ratio. *J. Mol. Catal. B: Enzym.* **2016**, *123*, 67–72. [CrossRef]

20. Chen, X.; Liu, Z.-Q.; Lin, C.-P.; Zheng, Y.-G. Efficient biosynthesis of ethyl (R)-4-chloro- 3-hydroxybutyrate using a stereoselective carbonyl reductase from *Burkholderia gladioli*. *BMC Biotechnol.* **2016**, *16*, 70. [CrossRef] [PubMed]
21. Nealon, C.M.; Musa, M.M.; Patel, J.M.; Phillips, R.S. Controlling substrate specificity and stereospecificity of alcohol dehydrogenases. *ACS Catal.* **2015**, *5*, 2100–2114. [CrossRef]
22. Stepankova, V.; Bidmanova, S.; Koudelakova, T.; Prokop, Z.; Chaloupkova, R.; Damborsky, J. Strategies for stabilization of enzymes in organic solvents. *ACS Catal.* **2013**, *3*, 2823–2836. [CrossRef]
23. Wang, Z.; Song, Q.; Yu, M.; Wang, Y.; Xiong, B.; Zhang, Y.; Zheng, J.; Ying, X. Characterization of a stereospecific acetoin(diacetyl) reductase from *Rhodococcus erythropolis* WZ010 and its application for the synthesis of (2S,3S)-2,3-butanediol. *Appl. Microbiol. Biotechnol.* **2014**, *98*, 641–650. [CrossRef] [PubMed]
24. Yu, M.; Huang, M.; Song, Q.; Shao, J.; Ying, X. Characterization of a (2R,3R)-2,3-butanediol dehydrogenase from *Rhodococcus erythropolis* WZ010. *Molecules* **2015**, *20*, 7156–7173. [CrossRef] [PubMed]
25. Abokitse, K.; Hummel, W. Cloning, sequence analysis, and heterologous expression of the gene encoding a (S)-specific alcohol dehydrogenase from *Rhodococcus erythropolis* DSM 43297. *Appl. Microbiol. Biotechnol.* **2003**, *62*, 380–386. [CrossRef] [PubMed]
26. Itoh, N.; Morihama, R.; Wang, J.; Okada, K.; Mizuguchi, N. Purification and characterization of phenylacetaldehyde reductase from a styrene-assimilating *Corynebacterium* strain, ST-10. *Appl. Environ. Microbiol.* **1997**, *63*, 3783–3788. [PubMed]
27. Kasprzak, J.; Bischoff, F.; Rauter, M.; Becher, K.; Baronian, K.; Bode, R.; Schauer, F.; Vorbrodt, H.-M.; Kunze, G. Synthesis of 1-(S)-phenylethanol and ethyl (R)-4-chloro-3-hydroxybutanoate using recombinant *Rhodococcus erythropolis* alcohol dehydrogenase produced by two yeast species. *Biochem. Eng. J.* **2016**, *106*, 107–117. [CrossRef]
28. Makino, Y.; Dairi, T.; Itoh, N. Engineering the phenylacetaldehyde reductase mutant for improved substrate conversion in the presence of concentrated 2-propanol. *Appl. Microbiol. Biotechnol.* **2007**, *77*, 833–843. [CrossRef] [PubMed]
29. Karabec, M.; Łyskowski, A.; Tauber, K.C.; Steinkellner, G.; Kroutil, W.; Grogan, G.; Gruber, K. Structural insights into substrate specificity and solvent tolerance in alcohol dehydrogenase ADH-A from *Rhodococcus ruber* DSM 44541. *Chem. Commun.* **2010**, *46*, 6314–6316. [CrossRef] [PubMed]
30. Ying, X.; Wang, Y.; Xiong, B.; Wu, T.; Xie, L.; Yu, M.; Wang, Z. Characterization of an allylic/benzyl alcohol dehydrogenase from *Yokenella* sp. strain WZY002, an organism potentially useful for the synthesis of α,β-unsaturated alcohols from allylic aldehydes and ketones. *Appl. Environ. Microbiol.* **2014**, *80*, 2399–2409. [CrossRef] [PubMed]
31. Martinez-Rojas, E.; Kurt, T.; Schmidt, U.; Meyer, V.; Garbe, L.-A. A bifunctional enzyme from *Rhodococcus erythropolis* exhibiting secondary alcohol dehydrogenase-catalase activities. *Appl. Microbiol. Biotechnol.* **2014**, *98*, 9249–9258. [CrossRef] [PubMed]
32. Kratzer, R.; Woodley, J.M.; Nidetzky, B. Rules for biocatalyst and reaction engineering to implement effective, NAD(P)H-dependent, whole cell bioreductions. *Biotechnol. Adv.* **2015**, *33*, 1641–1652. [CrossRef] [PubMed]
33. Glonke, S.; Sadowski, G.; Brandenbusch, C. Applied catastrophic phase inversion: A continuous non-centrifugal phase separation step in biphasic whole-cell biocatalysis. *J. Ind. Microbiol. Biotechnol.* **2016**, *43*, 1527–1535. [CrossRef] [PubMed]
34. Wei, L.; Zhang, M.; Zhang, X.; Xin, H.; Yang, H. Pickering emulsion as an efficient platform for enzymatic reactions without stirring. *ACS Sustain. Chem. Eng.* **2016**, *4*, 6838–6843. [CrossRef]
35. Yang, C.; Ying, X.; Yu, M.; Zhang, Y.; Xiong, B.; Song, Q.; Wang, Z. Towards the discovery of alcohol dehydrogenases: NAD(P)H fluorescence-based screening and characterization of the newly isolated *Rhodococcus erythropolis* WZ010 in the preparation of chiral aryl secondary alcohols. *J. Ind. Microbiol. Biotechnol.* **2012**, *39*, 1431–1443. [CrossRef] [PubMed]
36. Laemmli, U.K. Cleavage of structural proteins during the assembly of the head of bacteriophage T4. *Nature* **1970**, *227*, 680–685. [CrossRef] [PubMed]
37. Söding, J.; Biegert, A.; Lupas, A.N. The HHpred interactive server for protein homology detection and structure prediction. *Nucleic Acids Res.* **2005**, *33*, W244–W248. [CrossRef] [PubMed]
38. Schymkowitz, J.; Borg, J.; Stricher, F.; Nys, R.; Rousseau, F.; Serrano, L. The FoldX web server: An online force field. *Nucleic Acids Res.* **2005**, *33*, W382–W388. [CrossRef] [PubMed]

39. Trott, O.; Olson, A.J. AutoDock Vina: Improving the speed and accuracy of docking with a new scoring function, efficient optimization, and multithreading. *J. Comput. Chem.* **2010**, *31*, 455–461. [CrossRef] [PubMed]
40. Duan, Y.; Wu, C.; Chowdhury, S.; Lee, M.C.; Xiong, G.; Zhang, W.; Yang, R.; Cieplak, P.; Luo, R.; Lee, T.; et al. A point-charge force field for molecular mechanics simulations of proteins based on condensed-phase quantum mechanical calculations. *J. Comput. Chem.* **2003**, *24*, 1999–2012. [CrossRef] [PubMed]

Sample Availability: Samples of the compounds are not available from the authors.

© 2018 by the authors. Licensee MDPI, Basel, Switzerland. This article is an open access article distributed under the terms and conditions of the Creative Commons Attribution (CC BY) license (http://creativecommons.org/licenses/by/4.0/).

Article

Chemical Modification of Sweet Potato β-amylase by Mal-mPEG to Improve Its Enzymatic Characteristics

Xinhong Liang, Wanli Zhang, Junjian Ran, Junliang Sun *, Lingxia Jiao, Longfei Feng and Benguo Liu

School of Food Science, Henan Institute of Science and Technology, Xinxiang 453003, China; liangxinhong2005@163.com (X.L.); zwl6996468@126.com (W.Z.); ranjunjian@126.com (J.R.); jiaolingxia@163.com (L.J.); flf18738380903_007@163.com (L.F.); zzgclbg@126.com (B.L.)
* Correspondence: sjl@hist.edu.cn; Tel.: +86-135-0373-8786

Academic Editor: Stefano Serra
Received: 19 September 2018; Accepted: 23 October 2018; Published: 24 October 2018

Abstract: The sweet potato β-amylase (SPA) was modified by 6 types of methoxy polyethylene glycol to enhance its specific activity and thermal stability. The aims of the study were to select the optimum modifier, optimize the modification parameters, and further investigate the characterization of the modified SPA. The results showed that methoxy polyethylene glycol maleimide (molecular weight 5000, Mal-mPEG5000) was the optimum modifier of SPA; Under the optimal modification conditions, the specific activity of Mal-mPEG5000-SPA was 24.06% higher than that of the untreated SPA. Mal-mPEG5000-SPA was monomeric with a molecular weight of about 67 kDa by SDS-PAGE. The characteristics of Mal-mPEG5000-SPA were significantly improved. The K_m value, V_{max} and Ea in Mal-mPEG5000-SPA for sweet potato starch showed that Mal-mPEG5000-SPA had greater affinity for sweet potato starch and higher speed of hydrolysis than SPA. There was no significant difference of the metal ions' effect on Mal-mPEG5000-SPA and SPA.

Keywords: sweet potato β-amylase (SPA) 2; methoxy polyethylene glycol maleimide (Mal-mPEG) 3; chemical modification 4; enzymatic characteristics

---

1. Introduction

β-amylase (E.C. 3.2.1.2) is an exo-type saccharifying enzyme that can act on the α-1,4 glucosidic bonds and cleave off maltose units at the non-reducing end of starch molecules [1]. The cleave-off process is accompanied by Walden inversion that turns the product from α-maltose into β-maltose [2,3]. β-amylase exists widely in higher plants such as sweet potato, barley, wheat, and soybea [4], and is mainly applied in the industrial process including food, fermentation, textiles, pharmaceuticals, etc. [5,6]. Sweet potato β-amylase (SPA) is an important component of protein in sweet potato tubers next only to sporamin, and is primarily obtained by extraction and separation from the waste water of the sweet potato starch production [7]. As a bio-active biomacromolecule, SPA's biological activity and thermal stability are among the key factors that limit its application in the food industry.

Chemical modification of molecules is an effective means to increase enzymatic stability and biological activity. It can also effectively prolong the half-life of enzymes. Polyethylene glycol (PEG) is a good non-irritant amphipathic organic solvent without immunogenicity, antigenicity and toxicity [8,9]. The studies made by Abuehowski et al. as early as 1977 showed that proteins modified by PEG were of greater efficacy than the unmodified ones [10]. Meanwhile, through enzyme modification, the antigenicity of particular enzymes can be reduced or eliminated, and the enzymatic stability can be enhanced. Therefore, the techniques of PEG-modified proteins have been developing rapidly [11,12]. Presently, over 10 types of PEG-modified proteins have been certified by the US Food and Drug

Administration [13]. However, due to deficiencies such as frequent occurrence of crosslinking and agglomeration of -OH at both ends of PEG, its usage in protein modification is limited.

With similar properties to PEG, methoxy polyethylene glycol (mPEG) also works as a protein modifier with the active hydroxyl group at one end of PEG being blocked by the methoxy group. Modification through mPEG can change the relevant biological characteristics of enzymes or proteins, including hydrophobicity, surface charge, stability and water solubility [13–15]. It is reported that the activity, thermal stability and pH stability of enzymes modified by mPEG can be enhanced. Using mPEG to modify neutral protease changed the thermal stability of the enzyme, and the modified enzyme showed higher affinity for substrate [16]. Fang et al. [17] modified phospholipase C by using mPEG-succinimidyl succinate ester. The results indicated an increase by three times in the catalytic efficiency of the modified phospholipase C, and also higher thermal stability. Daba et al. [18] adopted glutaraldehyde (GA), mPEG chlorotriazine and trinitro-benzene-sulfonic acid (TNBS) to modify β-amylase in malt. The result indicated that mPEG chlorotriazine enhanced the enzymatic activity and thermal stability of β-amylase in malt. However, there are very few reports about mPEG modified SPA.

Methoxy polyethylene glycol N-hydroxylsuccinimide ester (NHS-mPEG5000, NHS-mPEG20000), Methoxy polyethylene glycol tosylate (Ts-mPEG5000, Ts-mPEG5000, Ts-mPEG10000, Ts-mPEG20000) and Methoxy polyethylene glycol maleimide (Mal-mPEG5000) were adopted in this study for chemical modification on SPA to screen the optimum modifier. The response surface method was applied to optimize the molar ratio of the optimum modifier to SPA, as well as the modification temperature, pH value and other parameters. The enzymatic properties of the modification enzyme under optimal parameters were studied to improve its catalytic activity and thermal stability, so as to help lay the foundation for its industrial application.

2. Results and Discussion

2.1. Screening of Modifiers

The change in SPA specific activity modified by 6 different modifiers was shown in Figure 1. Compared with the untreated SPA, Mal-mPEG5000, NHS-mPEG5000 and Ts-mPEG5000 could significantly increase SPA's specific activity ($p < 0.05$) in the same molar ratio. The specific activity of Mal-mPEG5000-SPA reached the maximum value at $(2.081 \pm 0.050) \times 10^4$ U/mg. There was a significant difference between Mal-mPEG5000 and other modifiers ($p < 0.05$). However, there was no significant difference between Ts-mPEG10000, Ts-mPEG20000 and NHS-mPEG20000 ($p > 0.05$). Therefore, Mal-mPEG5000 was selected as the optimal modifier for the next experiments.

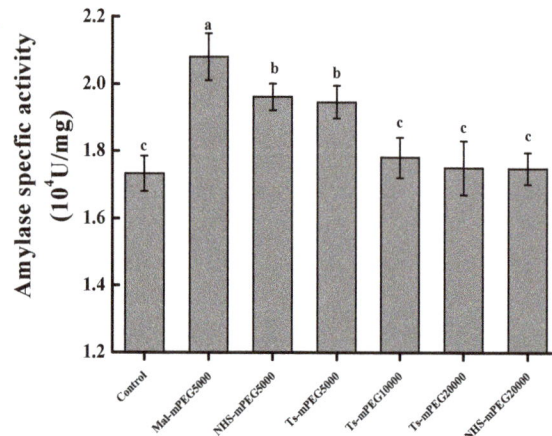

Figure 1. Effect of 6 types of modifiers on SPA activity. Significant difference in each column are expressed as different superscript letters ($p < 0.05$).

2.2. Effect of the Molar Ratio on Modification

The results from Figure 2 indicated that as the molar ratio rose from 1:1 to 1:4, increasingly high SPA specific activity was found along with the increase of Mal-mPEG5000 concentration. The enzymatic specific activity reached the maximum $(2.143 \pm 0.050) \times 10^4$ U/mg when the molar ratio was 1:4, an increase by 23.66% compared with that of the untreated SPA. As the concentration of Mal-mPEG5000 increased further, and the molar ratio increased to 1:6, the enzymatic specific activity decreased to $(2.016 \pm 0.051) \times 10^4$ U/mg, and there was a significant difference between the molar ratio of 1:4 and 1:6 ($p < 0.05$). Therefore, the optimal response molar ratio of SPA to Mal-mPEG5000 was concluded to be 1:4.

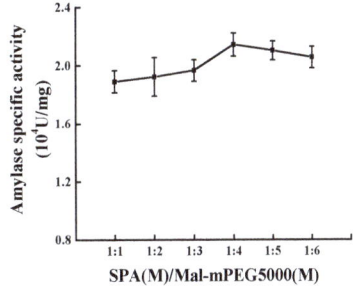

Figure 2. Effect of the molar ratio of SPA to Mal-mPEG5000 on SPA activity.

2.3. Effect of Temperature on Modification

The results from Figure 3 indicated that the Mal-mPEG5000-SPA specific activity gradually increased as the temperature rose from 25 °C to 55 °C. The specific activity reached the maximum $(2.131 \pm 0.059) \times 10^4$ U/mg at 55 °C. As the temperature increased to 65 °C, the specific activity dropped to $(1.801 \pm 0.055) \times 10^4$ U/mg, and decreased by 15.49%, and there was a significant difference between 55 °C and 65 °C ($p < 0.05$). Therefore, the optimal modification temperature was concluded to be 55 °C.

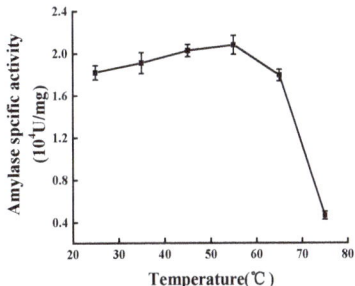

Figure 3. Effect of modification temperature on modification.

2.4. Effect of pH on Modification

The results from Figure 4 indicated that the Mal-mPEG5000-SPA specific activity gradually increased as the pH moved from 3 to 6. The specific activity reached the maximum $(2.155 \pm 0.046) \times 10^4$ U/mg at pH 6.0. As pH increased to 7.0, the specific activity was $(1.925 \pm 0.045) \times 10^4$ U/mg and decreased by 10.69%, and there was a significant difference between pH 6.0 and 7.0 ($p < 0.05$). Therefore, the optimal modification pH was concluded to be 6.0.

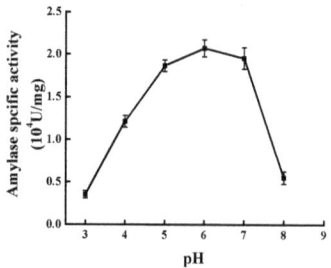

Figure 4. Effect of modification pH values on modification.

2.5. Effect of Time on Modification

The results from Figure 5 indicated that the Mal-mPEG5000-SPA specific activity increased as the modification time rose from 5 min to 10 min. The specific activity reached the maximum (2.142 ± 0.059) × 10^4 U/mg at 10 min. As the modification time further increased, no significant change was noted in the enzymatic specific activity ($p = 0.05$). The ANVOA results showed no significant difference of enzymatic specific activity between 10 min and 30 min. Therefore, taking into account of practical application, the optimal modification time was concluded to be 10 min.

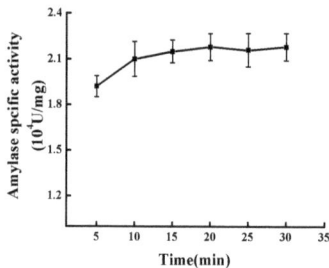

Figure 5. Effect of modification time on modification.

2.6. Optimizing the Modification Procedure

According to the effect of the molar ratio, modification temperature, pH and time, the time has little effect on modification, and modification pH (A), the molar ratio (B) and modification temperature (C) were selected as three factors to optimize the modification procedure by central composite design (CCD). With modification time set at 10 min, and Mal-mPEG5000-SPA specific activity (Y) as the response value, Box-Behnken design principles were followed to conduct a test of three factors and three levels. Results were shown in Table 1.

Table 1. Experimental design and results for the activity of Mal-mPEG5000-SPA.

Runs	Coded Levels			Response Value
	pH	Molar Ratio	Temperature (°C)	(×10^4 U/mg)
1	−1	−1	0	1.7763
2	0	−1	−1	1.8858
3	0	0	0	2.1592
4	0	0	0	2.1481
5	1	0	−1	1.8142
6	0	1	−1	1.8667
7	0	−1	1	1.8758
8	1	−1	0	1.9097

Table 1. Cont.

Runs	Coded Levels			Response Value
	pH	Molar Ratio	Temperature (°C)	($\times 10^4$ U/mg)
9	0	0	0	2.1323
10	−1	1	0	1.8620
11	−1	0	1	1.8226
12	0	0	0	2.1597
13	0	0	0	2.1326
14	0	1	1	1.9097
15	1	0	1	1.9097
16	1	1	0	1.8619
17	−1	0	−1	1.8362

Multiple regression analysis was adopted to the experimental data by using the software Design-Expert V8.0.6. The response, Y (Mal-mPEG5000-SPA specific activity), was selected as the test variables, and by the second order, a polynomial equation was developed (Equation (1)).

$$Y = 2.15 + 0.025A + 0.00528B + 0.016C - 0.033AB + 0.027AC + 0.0011BC - 0.17A^2 - 0.13B^2 - 0.13C^2 \quad (1)$$

The analysis of variance (ANOVA) for the response Y, the Mal-mPEG5000-SPA specific activity, was shown in Table 2. The regression model was highly significant ($p < 0.01$), while the lack of fit was not significant ($p = 0.317 > 0.05$) and the value of the determination coefficient (R^2) was 0.996, which indicated the goodness of fit of the regression model [19]. Based on the analysis of experimental data in Tables 1 and 2, the regression model demonstrated a high correlation, and can be used as theoretical prediction for the enzymatic specific activity of Mal-mPEG5000 modified SPA. The significance test on the regression model suggested that the effect on enzymatic specific activity was as follows: Modification pH > modification temperature > Molar ratio of SPA to Mal-mPEG5000.

Table 2. Analysis of variance (ANOVA) for the experimental results.

Source	Sum of Squares	df	Mean Square	F-Value	p-Value	Significance
Model	0.3	9	0.034	145.34	<0.0001	**
A	4.91×10^{-3}	1	4.91×10^{-3}	21.05	0.0025	**
B	2.23×10^{-4}	1	2.23×10^{-4}	0.96	0.3606	
C	1.95×10^{-3}	1	1.95×10^{-3}	8.36	0.0233	*
AB	4.48×10^{-3}	1	4.48×10^{-3}	19.23	0.0032	**
AC	2.98×10^{-3}	1	2.98×10^{-3}	12.76	0.0091	**
BC	4.62×10^{-4}	1	4.62×10^{-4}	1.98	0.2022	
A2	0.12	1	0.12	507.79	<0.0001	**
B2	0.067	1	0.067	288.2	<0.0001	**
C2	0.074	1	0.074	319.52	<0.0001	**
Residual	1.63×10^{-3}	7	2.33×10^{-4}			
Lack of fit	8.97×10^{-4}	3	2.99×10^{-4}	1.63	0.317	
Pure error	7.35×10^{-4}	4	1.84×10^{-4}			
Cor total	0.31	16				
R2	0.996					

* indicate significant ($p < 0.05$); ** indicate highly significant ($p < 0.01$).

In order to further comprehend the interaction between the parameters, the response surfaces were obtained using Equation (2), which were plotted between two independent variables and the other independent variable was set at the zero-coded level. The analysis on the response was shown in Figure 6a–c.

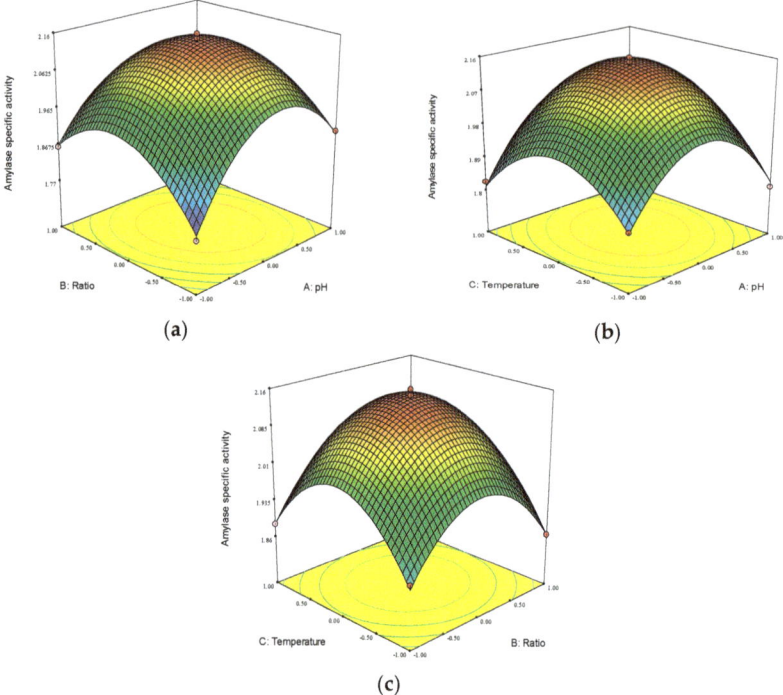

Figure 6. Response surface plot showing the effects of the variables on the activity of Mal-mPEG5000-SPA. The three independent variables set were modification pH (**a**), the molar ratio of SPA to Mal-mPEG5000 (**b**) and modification temperature (**c**).

Figure 6a indicated that with the increase of the modification pH and the molar ratio, the Mal-mPEG5000-SPA specific activity first increased, and then declined, meanwhile, steep response surfaces and oval contour line were found, which indicated significant interaction of the pH value and the molar ratio.

Figure 6b suggested that with the increase of modification pH and temperature, the Mal-mPEG5000-SPA specific activity firstly increased, and then decreased, and the oval contour line was noticed, which demonstrated a significant interaction of modification pH value temperature.

Figure 6c revealed that with the increase of the molar ratio and modification temperature, the Mal-mPEG5000-SPA specific activity increased firstly before it declined, and the occurrence of circular contour line showed no significant interaction of the molar ratio and modification temperature.

Through the analysis by the software Design-Expert, the optimized combination for maximum Mal-mPEG5000-SPA specific activity was determined through canonical analysis of the response surfaces and contour plots as A = 6.08, B = 4.04 and C = 58.85 °C, i.e., the predicted value of Mal-mPEG5000-SPA specific activity of 2.247×10^4 U/mg. under modification pH 6.08, the molar ratio 1:4.04, and modification temperature 58.85 °C. Considering the practicality of the validation test, the optimum parameters were corrected into modification pH 6.0, the molar ratio 1:4, and modification temperature 58 °C. The specific activity of Mal-mPEG5000-SPA was determined under the corrected parameters to be $(2.150 \pm 0.055) \times 10^4$ U/mg, an increase by 24.06% than that of the untreated one. The validation test showed that the optimization results from the response surface method were reliable, and adopting response surface method to optimize the modification process of Mal-mPEG5000 on SPA was feasible.

2.7. Separation and Purification of Mal-mPEG5000-SPA

Mal-mPEG5000-SPA was separated and purified by AKTA purifier^{TM}10 with Superdex^{TM}75 gel column (Figure 7a–c). Separated by Superdex^{TM}75 gel column, there were three protein peaks: S_1, S_2 and S_3 (Figure 7a), and the eluents in the collection tubes corresponding to the peak position were collected and determined for enzymatic specific activity. The specific activity of S_1 and S_2 was $(1.245 \pm 0.047) \times 10^4$ U/mg and $(0.533 \pm 0.036) \times 10^4$ U/mg respectively as shown in Figure 7c. As the peak of SPA elution was noted around 17 min according to the result of the pre-test, the elution peak S_1 was collected, centrifuged and concentrated for further purification.

Mal-mPEG5000-SPA was separated for the second time using Superdex^{TM}75 gel column, and there were elution peaks S_{11} and S_{12} (Figure 7b). The specific activity of S_{11} and S_{12} was determined to be $(1.422 \pm 0.057) \times 10^4$ U/mg and $(0.061 \pm 0.009) \times 10^4$ U/mg respectively (Figure 7d). According to the separation principles of gel column chromatography, different positions of the peaks suggest different molecular weight of the protein under each peak. The position of SPA peak was around 17 min, while the position of Mal-mPEG5000-SPA was about 7 min, which indicated change in molecular weight as SPA was modified by Mal-mPEG5000.

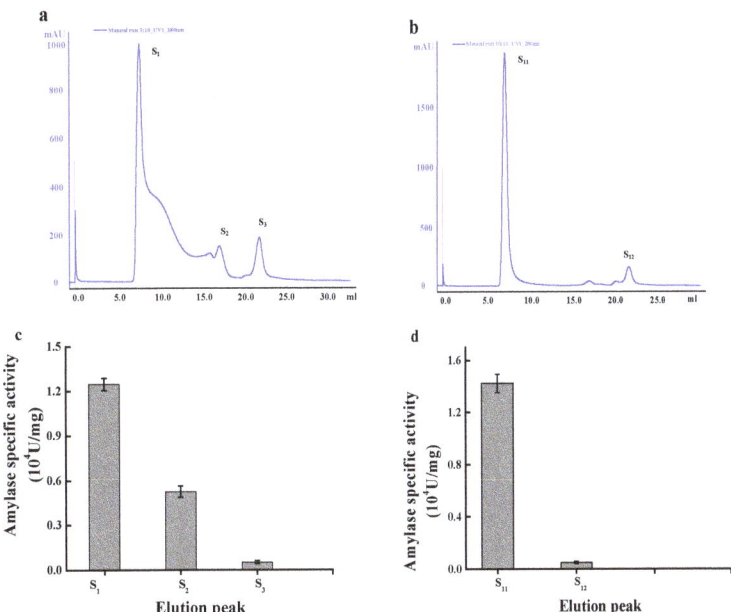

Figure 7. Separation curve of Mal-mPEG5000-SPA by gel column Superdex^{TM}75 (**a**) Elution profile of Mal-mPEG5000-SPA by Superdex^{TM}75 for the first time; (**b**) Elution profile of Mal-mPEG5000-SPA by Superdex^{TM}75 for the second time; (**c**) The specific activity of the elution fractions from Superdex^{TM}75 for the first time; (**d**) The specific activity of the elution fractions from Superdex^{TM}75 for the second time.

SDS-PAGE gel electrophoresis was carried out on the collected elution peaks S_1 and S_{11}. The results were shown in Figure 8a,b. An obvious band of S_1 was found above the SPA band as indicated in Figure 8a. The band was of Mal-mPEG5000-SPA, whose molecular weight was about 67 kDa compared with the standard marker. Figure 8b demonstrated the electrophoretogram of elution peak S_{11}. A single band was noted after separation twice by Superdex^{TM}75, and Mal-mPEG5000-SPA was obtained through purification.

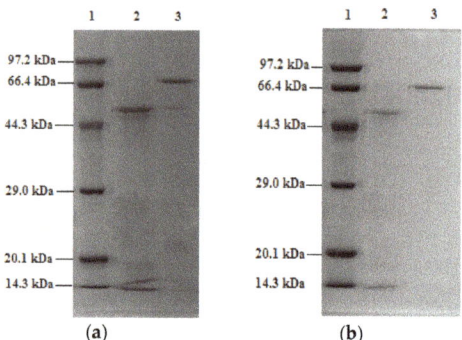

Figure 8. SDS-PAGE electrophoresis spectra of Mal-mPEG5000-β-SPA (**a**) separation by Superdex™75 for the first time; Lane 1: marker proteins; Lane 2: SPA; Lane 3: Mal-mPEG5000-SPA by Superdex™75. (**b**) separation by Superdex™75 for the second time; Lane 1: marker proteins; Lane 2: SPA; Lane 3: Mal-mPEG5000-SPA by Superdex™75.

2.8. Optimum Temperature and Thermal Stability of Mal-mPEG5000-SPA and SPA

According to Figure 9a, the maximum specific activity of SPA was found to be $(1.733 \pm 0.050) \times 10^4$ U/mg at 55 °C, while the maximum of Mal-mPEG5000-SPA was $(2.121 \pm 0.058) \times 10^4$ U/mg at 45 °C, an increase by 22.38% compared to the untreated one. Moreover, the activities of Mal-mPEG5000-SPA at 50 °C, 55 °C and 60 °C were determined as $(2.074 \pm 0.060) \times 10^4$ U/mg, $(2.057 \pm 0.062) \times 10^4$ U/mg, and $(2.031 \pm 0.064) \times 10^4$ U/mg respectively, all significantly higher than that of untreated SPA ($p < 0.05$). These results indicated that the enzymatic specific activity of modification enzyme was significantly improved after Mal-mPEG5000 modification.

Figure 9b showed the thermal stability of Mal-mPEG5000-SPA and SPA. No significant difference in SPA specific activity at 20 °C and 25 °C for 1 h was noted. However, when the temperature rose to 30 °C, the specific activity of SPA was significantly reduced ($p < 0.05$), which indicated high sensitiveness of SPA to temperature. No significant difference in the specific activity of Mal-mPEG5000-SPA from 20 °C to 45 °C for 1 h was found. But when the temperature rose to 50 °C, the specific activity of Mal-mPEG5000-SPA started to decline ($p < 0.05$), which suggested significant improvement of the thermal resistance of Mal-mPEG5000-SPA. These results showed that hydrolyzed starch by Mal-mPEG5000-SPA has a wider range of application in the food industry, which can significantly enhance enzymatic efficiency to reduce production costs.

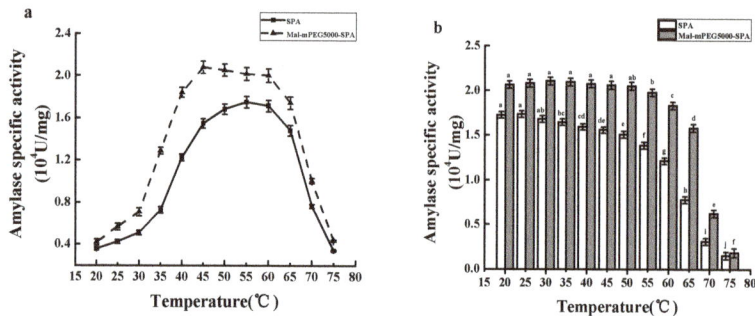

Figure 9. Temperature profiles of Mal-mPEG5000- SPA and SPA. (**a**) Optimum enzymolysis temperature; (**b**) Thermal stability. Significant difference in each column are expressed as different superscript letters ($p < 0.05$).

2.9. Optimum pH and pH Stability of Mal-mPEG5000-SPA and SPA

Figure 10a showed that Mal-mPEG5000-SPA and SPA both showed relatively high specific activity from pH 5.0 to 7.0. At pH 6.0, the specific activity of Mal-mPEG5000-SPA and SPA reached respective maximum values at $(2.061 \pm 0.051) \times 10^4$ U/mg and $(1.733 \pm 0.050) \times 10^4$ U/mg, and the specific activity was improved by 18.92% after modification.

Figure 10b showed the pH stability of Mal-mPEG5000-SPA and SPA. A significant difference in SPA's specific activity at pH 6.0 and 6.5 was found ($p < 0.05$), while the difference for Mal-mPEG5000-SPA was not significant ($p = 0.05$), which indicated higher adaptability of Mal-mPEG5000-SPA to a pH environment. In addition, the specific activity of Mal-mPEG5000-SPA was constantly higher than that of SPA under pH value from 4.0 to 7.5, which indicated lower susceptibility of Mal-mPEG5000-SPA to pH environment. This means Mal-mPEG5000-SPA could applied more broadly and is more suitable for industrial application.

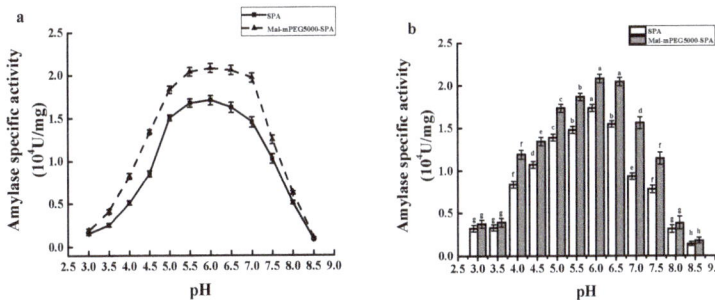

Figure 10. pH profiles of Mal-mPEG5000- SPA and SPA. (**a**) Optimum enzymolysis pH; (**b**) pH stability. Significant difference in each column are expressed as different superscript letters ($p < 0.05$).

2.10. Kinetic Parameters of Mal-mPEG5000-SPA

The results of kinetic parameters of Mal-mPEG5000-SPA were shown in Table 3, the K_m value for sweet potato starch, potato starch, corn starch, soluble starch, amylase and amylopectin by Mal-mPEG5000-SPA hydrolysis was respectively (1.63 ± 0.033) mg/mL, (2.06 ± 0.028) mg/mL, (2.36 ± 0.063) mg/mL, (1.84 ± 0.025) mg/mL, (2.18 ± 0.029) mg/mL and (2.11 ± 0.052) mg/mL. The V_{max} value was respectively (32.06 ± 0.61) mmol/min/mL, (16.23 ± 0.32) mmol/min/mL, (10.66 ± 0.37) mmol/min/mL, (20.88 ± 0.78) mmol/min/mL, (13.35 ± 0.38) mmol/min/mL, (13.89 ± 0.41) mmol/min/mL. The E_a value was respectively (11.07 ± 0.43) kJ/mol, (18.24 ± 1.12) kJ/mol, (26.52 ± 1.21) kJ/mol, (14.71 ± 1.15) kJ/mol, (21.16 ± 1.12) kJ/mol and (21.66 ± 0.8) kJ/mol. These results indicated that Mal-mPEG5000-SPA had the lowest Michaelis constant for sweet potato starch. Therefore, Mal-mPEG5000-SPA was determined to have the strongest binding affinity for sweet potato starch. Next only to sweet potato starch, soluble starch was followed by potato starch, amylase and amylopectin. Compared with the SPA kinetic parameters reported in earlier studies [20], for sweet potato starch hydrolyzed by Mal-mPEG5000-SPA, K_m declined by 12.95%, V_{max} increased by 26.87%, and Ea dropped by 12.63%, which suggested Mal-mPEG5000-SPA showed stronger affinity for sweet potato starch than SPA.

Table 3. Kinetic and activation energy parameters of Mal-mPEG5000-SPA.

Substrates	K_m (mg/mL)	V_{max} (mmol/min/mL)	E_a (kJ/mol)
Sweet potato starch	1.87 ± 0.032 [a]	19.32 ± 0.65 [e]	12.67 ± 0.73 [a]
Potato starch	2.16 ± 0.031 [c]	14.62 ± 0.35 [c]	19.84 ± 1.03 [c]
Corn starch	2.56 ± 0.053 [e]	9.36 ± 0.27 [a]	28.23 ± 1.37 [e]

Table 3. Cont.

Substrates	K_m (mg/mL)	V_{max} (mmol/min/mL)	E_a (kJ/mol)
Soluble starch	1.96 ± 0.029 [b]	16.75 ± 0.68 [d]	16.61 ± 0.99 [b]
Amylase	2.32 ± 0.037 [d]	12.85 ± 0.42 [b]	22.26 ± 1.09 [d]
Amylopectin	2.37 ± 0.045 [d]	12.53 ± 0.41 [b]	22.54 ± 0.98 [d]

Significant difference in each column are expressed as different superscript letters ($p < 0.05$).

2.11. Effect of Metal Ions on the Activity of Mal-mPEG5000-SPA

As shown in Table 4, the relative specific activity after addition of Mn^{2+}, K^+, Zn^{2+} and Ca^{2+} was respectively (143.48 ± 6.25)%, (115.49 ± 4.51)%, (111.28 ± 4.82)% and (102.88 ± 4.32)%, which demonstrated that Mn^{2+}, K^+, Zn^{2+} and Ca^{2+} had good activation effect. Mg^{2+} showed a slight inhibitive effect on Mal-mPEG5000-SPA, with relative specific activity standing at (95.25 ± 4.14)%. However, Cu^{2+}, NH_4^+, Fe^{3+}, Al^{3+}, Ba^{2+} and EDTA demonstrated a relatively strong inhibitory effect on Mal-mPEG5000-SPA, with relative specific activity reduced to 55–80%. Meanwhile, under the effect of Cd^{2+}, Hg^+ and Ag^+, the relative specific activity were all below 30%, among which Hg^+ and Ag^+ indicated the strongest inhibitory effect on Mal-mPEG5000-SPA. Compared with the effect of metal ions on SPA reported in earlier studies [21], the effect of metal ions on Mal-mPEG5000-SPA and on SPA showed no significant difference ($p = 0.05$).

Table 4. Effect of metal ions on the activity of Mal-mPEG5000-SPA.

Metal Ions	Relative Activity (%)
Control	100
Ca^{2+}	102.88 ± 4.32 [c]
Mg^{2+}	95.25 ± 4.14 [d]
Cu^{2+}	60.95 ± 3.61 [f]
Zn^{2+}	111.28 ± 4.82 [b]
Mn^{2+}	143.48 ± 6.25 [a]
K^+	115.49 ± 4.51 [b]
NH_4^+	78.89 ± 4.05 [e]
Hg^+	5.21 ± 0.63 [i]
Ag^+	3.28 ± 0.12 [i]
Al^{3+}	45.81 ± 3.52 [g]
Fe^{3+}	80.85 ± 4.62 [e]
Ba^{2+}	59.68 ± 3.92 [f]
Cd^{2+}	22.85 ± 2.06 [h]
EDTA	55.84 ± 3.27 [f]

Significant difference in each column are expressed as different superscript letters ($p < 0.05$).

3. Materials and Methods

3.1. Materials

Sweet potato, molar weight 56.043 kDa, harvested in Xinxiang City, Henan Province, China, its cultivar was Shangshu 19. NHS-mPEG5000, NHS-mPEG20000, Ts-mPEG5000, Ts-mPEG10000, Ts-mPEG20000 and Mal-mPEG5000 were from Nanocs (New York, NY, USA; purity ≥ 95%). All other reagents were of analytical reagent grade and used without further purification.

3.2. Separation and Purification of SPA

Using the methods described by Liang et al. [21] with minor modification. SPA was separated and purified with the following steps. Fresh sweet potatoes were washed and sliced into chips. 100 g sweet potatoes were weighed out, and 200 mL distilled water was added for pulverization in a pulverizer

(SQ2002, Shanghai Shuaijia Electronic Technology Company, China) for 1 min. The sample was filtered through 40 mesh sieve, and the resulting filtrate was placed in a refrigerated centrifuge at 4 °C for centrifugation at 4000 rpm for 15 min. The supernatant was obtained, and ammonium sulfate was added to the supernatant to 70% saturation. The resulting solution was stored in a refrigerator at 4 °C for 4 h. The centrifuge parameters were set at 4 °C and 8000 rpm for refrigerated centrifugation for 15 min. The resulting precipitate was collected, and dissolved in a bufferA. Ammonium sulfate was desalinated by using 1 kDa ultrafiltration membrane for 4 h. Protein purifier (AKTA purifier™10, General Electric Company, Boston, Massachusetts, MA, USA) was used to purify the enzyme, and Mono Q anion exchange chromatographic column and Superdex™75 gel column were adopted for separation and purification with detection wavelength set at 280 nm. The buffers for purification were as follows. Buffer A: 20 mM pH 5.8 disodium-hydrogen phosphate-citric acid, Buffer B: 1 mol/L NaCl. The buffers were filtered through a 0.22 μm membrane, ultrasound treated for 20 min, and stored in a refrigerator at 4 °C. Buffer A was used to equilibrate the chromatographic column. The sample was injected after equilibration of buffer to the baseline. The column was washed with 5 column volumes of buffer A, and then eluted with a gradient from 0–100% Buffer B at the flow rate of 1.0 mL/min, and maximum back-pressure of 4 MPa for 30 min. A collector was used to obtain 1 mL fractions. After the elution was complete, the enzymes in the collection tube corresponding to the peak position were collected, frozen and dried in a vacuum freeze dryer (Thermo Savan, Thermo Electron Co., Waltham, MA, USA) and stored at 4 °C for later usage. The separation and purification of mPEG5000-β-SPA: 70% ammonium sulfate was added into the reaction liquid to precipitate. The rest steps were the same as those of the β-amylase separation and purification.

The molecular mass and the purity of the enzyme was conducted by SDS-PAGE according to the Laemmli [22] method, and 15% (w/v) polyacrylamide gel was adopted.

3.3. Protein Content and Enzyme Assay

Protein content was measured as described by Lowry et al. [23], and bovine serum albumin was used as the standard. Enzyme assay was determined by the dinitrosalicylic acid (DNS) method according to Sagu et al. [24] with minor modification. The enzyme assay was performed at 40 °C, pH 5.8, and 1 mg maltose released per hour from 1.1% soluble starch was defined as a unit of enzyme specific activity. In this paper, the enzyme activity was indicated by specific enzyme activity, expressed as U/mg.

3.4. Screening of Modifiers

The molar ratio of SPA to modifiers was determined at 1:4. SPA and 6 different types of modifiers (NHS-mPEG5000, NHS-mPEG20000, Ts-mPEG5000, Ts-mPEG10000, Ts-mPEG20000 and Mal-mPEG5000) were placed in the buffer (disodium-hydrogen phosphate-citric acid, pH 6.0), respectively, and then kept in a water-bath thermostatic metal oscillator at 55 °C for 10 min. The reaction mixture was obtained for dialysis. After the dialysis was complete, the enzyme specific activity was measured to obtain the optimal modifier.

3.5. Selection of Relevant Variables and Experimental Design

SPA was modified by Mal-mPEG5000 at the molar ratio of SPA to Mal-mPEG5000 from 1:1 to 1:6, for modification temperature from 25 °C to 75 °C and for modification pH from 3.0 to 8.0. The enzyme specific activity of modified enzyme (expressed as Mal-mPEG5000- SPA) was used as the index for evaluating the effects of the different modification parameters to be optimized via response surface methodology (RSM). A central composite design (CCD) were used with three variables and three levels for optimizing the modification conditions. These three factors were modification pH (A), the molar ratio of SPA to Mal-mPEG5000 (B) and modification temperature (C) (Table 5). A second-order polynomial equation (Equation (2)) for the variables was developed:

$$Y = \alpha_0 + \alpha_1 A + \alpha_2 B + \alpha_3 C + \alpha_{11} A^2 + \alpha_{22} B^2 + \alpha_{33} C^2 + \alpha_{12} AB + \alpha_{13} AC + \alpha_{23} BC \qquad (2)$$

Y is the predicted response, α_0 is the intercept; α_1, α_2, α_3, linear coefficients; α_{11}, α_{22}, α_{33}, squared coefficients; and α_{12}, α_{13}, α_{23}, interaction coefficients. Analysis of variance was used to evaluate the model's adequacy and to determine the regression coefficients and their statistical significance. The response surface contour plots showed how the independent variables interacted and how those interactions influenced the overall response [20].

Table 5. Independent variables and levels for central composite design (CCD).

Factors	Levels		
	−1	0	1
Modification pH	5.0	6.0	7.0
Molar ratio of SPA to Mal-mPEG5000	1:3	1:4	1:5
Temperature/°C	45	55	65

3.6. Enzyme Characterization

Influence of temperature on Mal-mPEG5000-SPA and SPA: The optimum temperature of Mal-mPEG5000-SPA was investigated at pH 5.8 over a temperature range of 20–75 °C. SPA and Mal-mPEG5000-SPA was incubated in 20 mM disodium-hydrogen phosphate-citric acid buffer (pH 5.8) during 60 min, respectively. The thermostability of the enzyme residual specific activity were determined under different temperature conditions (20–75 °C).

Influence of pH on Mal-mPEG5000-SPA and SPA: The optimum pH of Mal-mPEG5000-SPA was investigated at 50 °C over a pH range of 3.0–8.5. Mal-mPEG5000-SPA was preincubated for 60 min under 50 °C, the residual specific activity was determined at a pH range of 3.0–8.5 to evaluate the pH stability.

3.7. Effect of Metal Ions on Mal-mPEG5000-SPA and SPA

Incubation of the enzyme was carried out by 10 mM of metal ions salts (chlorides including Ca^{2+}, Mg^{2+}, Cu^{2+}, Zn^{2+}, Mn^{2+}, K^+, NH_4^+, Hg^+, Al^{3+}, Fe^{3+}, Ba^{2+} and Cd^+), $AgNO_3$ and EDTA at 40 °C for 30 min, and the residual activities were determined respectively. An enzyme that did not contain metal ions was chosen as control (100%).

3.8. Kinetic Constant

The Michaelis constant (K_m) and the maximum velocity (V_{max}) of Mal-mPEG5000-SPA and SPA was defined by Lineweaver-Burk plot. Different concentrations of substrates (1–20 mg/mL) in 20 mM disodium-hydrogen phosphate-citric acid buffer pH 5.8 were used and specific activity was assessed by the DNS method. The values of K_m and V_{max} were calculated respectively based on the double reciprocal plot [25].

3.9. Activation Energy (Ea)

E_a of the enzyme was measured by Arrhenius equation ($lnk_{cat} = lnk_0 - Ea/RT$), and the temperature was set from 25 °C to 65 °C. Representing 1/T in K on the axis of x and natural log of specific activity on the axis of y, Arrhenius plot was generated. E_a was defined based on the value of the slope.

3.10. Statistical Analysis

All experiments were carried out in triplicate. SPSS 22.0 statistical software (SPSS Inc., Chicago, IL, USA) was used for the analysis of variance, and the significance test ($p < 0.05$) was performed by Duncan new complex method. All analyses were done in triplicate, and the statistical significance was determined using the mean values ± standard deviation.

4. Conclusions

In this study, we selected Mal-mPEG5000 as the best modifier from six modifiers and optimized the modification parameters by the response surface method. We used column chromatography to isolate and purify Mal-mPEG5000-SPA from the reaction solution between SPA and Mal-mPEG5000. Thereafter, the enzymatic properties were determined and compared in the presence and absence of Mal-mPEG5000. For the results, under the optimal conditions (the molar ratio of Mal-mPEG5000 to SPA, 1:4, modification temperature, 58 °C, pH 6.0), the specific activity of Mal-mPEG5000-SPA was $(2.150 \pm 0.055) \times 10^4$ U/mg, an increase by 24.06% than that of the unmodified SPA. Mal-mPEG5000-SPA was separated and purified, and a single band was noticed and its molecular weight was about 67 kDa. The enzyme properties of Mal-mPEG5000-SPA were significantly improved as the optimum temperature declined from 55 °C to 45 °C, the thermal stability and pH stability were obviously enhanced; the K_m value of sweet potato starch by Mal-mPEG5000-SPA declined by 12.95%, while V_{max} increased by 26.87%, and Ea dropped by 12.63%, which showed that Mal-mPEG5000-SPA had greater affinity for sweet potato starch and higher the speed of hydrolysis than SPA; as for Mal-mPEG5000-SPA, Mn^{2+}, K^+, Zn^{2+} and Ca^{2+} demonstrated activation effect; Mg^{2+} showed a slight inhibitory effect; Cu^{2+}, NH_4^+, Fe^{3+}, Al^{3+}, Ba^{2+}, EDTA, Hg^+ and Ag^+ had a relatively strong inhibitory effect; there was no significant difference of metal ions' effect on Mal-mPEG5000-SPA and SPA. It is concluded that Mal-mPEG5000-SPA will have a wider range of application in beer processing, maltose production and other food industries. Therefore, the studies on SPA application in food industries are theoretically and practically significant.

Author Contributions: J.S. and X.L. conceived and designed the experiments; X.L. and W.Z. performed the experiments; J.R. and L.J. analyzed the data; L.F. and B.L. provided the software and formal analysis.

Funding: This research was supported by the National Natural Science Foundation of China (No. 31771941), the Program for Science & Technology Innovation Talents in Universities of Henan Province (No. 16IRTSTHN007) and the Major Science and Technology Project in Henan (No. 182102110109).

Conflicts of Interest: The authors declare no conflict of interest.

References

1. Chang, C.T.; Lion, H.Y.; Tang, H.L.; Sung, H.Y. Activation, purification and properties of beta-amylase from sweet potatoes (Ipomoea batatas). *Biotechnol. Appl. Biochem.* **1996**, *126*, 120–128.
2. Dicko, M.H.; Searle-van Leeuwen, M.J.F.; Beldman, G.; Ouedraogo, O.G.; Hilhorst, O.G.; Traore, A.S. Purification and characterization of β-amylase from Curculigo pilosa. *Appl. Microbiol. Biotechnol.* **1999**, *52*, 802–805. [CrossRef]
3. Kaplan, F.; Dong, Y.S.; Guy, C.L. Roles of β-amylase and starch breakdown during temperatures stress. *Physiol. Plant.* **2010**, *126*, 120–128. [CrossRef]
4. Vajravijayan, S.; Pletnev, S.; Mani, N.; Nandhagopal, N.; Gunasekaran, K. Structural insights on starch hydrolysis by plant β-amylase and its evolutionary relationship with bacterial enzymes. *Int. J. Biol. Macromol.* **2018**, *113*, 329–337. [CrossRef] [PubMed]
5. Miao, M.; Li, R.; Huang, C.; Jiang, B.; Zhang, T. Impact of beta-amylase degradation on properties of sugary maize soluble starch particles. *Food Chem.* **2015**, *177*, 1–7. [CrossRef] [PubMed]
6. Mihajlovski, K.R.; Radovanovic, N.R.; Veljovic, D.N.; Siler-Marinkovic, S.S.; Dimitrijevic-Brankovic, S.I. Improved β-amylase production on molasses and sugar beet pulp by a novel strain Paenibacillus chitinolyticus CKS1. *Ind. Crops Prod.* **2016**, *80*, 115–122. [CrossRef]

7. Li, H.S; Oba, K. Major Soluble Proteins of Sweet Potato Roots and Changes in Proteins after Cutting, Infection, or Storage. *J. Agric. Chem. Soc. Jpn.* **1985**, *49*, 737–744.
8. Li, L.; Shang, B.; Hu, L.; Shao, R.; Zhen, Y. Site-specific PEGylation of lidamycin and its antitumor activity. *Acta Pharm. Sin. B* **2015**, *5*, 264–269. [CrossRef] [PubMed]
9. Hsieh, Y.P.; Lin, S.C. Effect of PEGylation on the Activity and Stability of Horseradish Peroxidase and L-N-Carbamoylase in Aqueous Phases. *Process Biochem.* **2015**, *50*, 1372–1378. [CrossRef]
10. Abuchowski, A.; van Es, T.; Palczuk, N.C.; Davis, F.F. Alteration of immunological properties of bovine serum albumin by covalent attachment of polyethylene glycol. *J. Biol. Chem.* **1977**, *252*, 3578–3581. [PubMed]
11. Pfister, D.; Morbidelli, M. Process for protein PEGylation. *J. Control. Release Off. J. Control. Release Soc.* **2014**, *180*, 134–149. [CrossRef] [PubMed]
12. Roberts, M.J.; Bentley, M.D.; Harris, J.M. Chemistry for peptide and protein PEGylation. *Adv. Drug Deliv. Rev.* **2002**, *54*, 459–476. [CrossRef]
13. Dozier, J.K.; Distefano, M.D. Site-Specific PEGylation of Therapeutic Proteins. *Int. J. Mol. Sci.* **2015**, *16*, 25831–25864. [CrossRef] [PubMed]
14. Maiser, B.; Baumgartner, K.; Dismer, F.; Hubbuch, J. Effect of lysozyme solid-phase PEGylation on reaction kinetics and isoform distribution. *J. Chromatogr. B* **2015**, *1002*, 313–318. [CrossRef] [PubMed]
15. Xu, Y.; Shi, Y.; Zhou, J.; Yang, W.; Bai, L.; Wang, S.; Jin, X.; Niu, Q.; Huang, A.; Wang, D. Structure-based antigenic epitope and PEGylation improve the efficacy of staphylokinase. *Microb. Cell Fact* **2017**, *16*, 197. [CrossRef] [PubMed]
16. Zhao, S.G.; Fang, L.M.; Wang, L.; Yin, R.C. Preparation and properties of neutral protease modified with monomethoxypolyethylene glycol. *Food Ferment. Ind.* **2014**, *40*, 160–163.
17. Fang, X.; Wang, X.; Li, G.; Zeng, J.; Li, J.; Liu, J. SS-mPEG chemical modification of recombinant phospholipase C for enhanced thermal stability and catalytic efficiency. *Int. J. Biol. Macromol.* **2018**, *111*, 1032–1039. [CrossRef] [PubMed]
18. Daba, T.; Kojima, K.; Inouye, K. Chemical modification of wheat beta-amylase by trinitrobenzenesulfonic acid, methoxypolyethylene glycol, and glutaraldehyde to improve its thermal stability and activity. *Enzyme Microb. Technol.* **2013**, *53*, 420–426. [CrossRef] [PubMed]
19. Bhardwaj, S.K.; Basu, T. Study on binding phenomenon of lipase enzyme with tributyrin on the surface of graphene oxide array using surface plasmon resonance. *Thin Solid Films* **2018**, *645*, 10–18. [CrossRef]
20. Liang, X.; Ran, J.; Sun, J.; Wang, T.; Jiao, Z.; He, H.; Zhu, M. Steam-explosion-modified optimization of soluble dietary fiber extraction from apple pomace using response surface methodology. *CyTA-J. Food* **2018**, *16*, 1–7. [CrossRef]
21. Liang, X.; Zhang, W.; Wang, Y.; Sun, J. Purification and characterization of β-amylase from sweet potato (Baizhengshu 2) tuberous roots. *Res. J. Biotechnol.* **2018**, *13*, 84–91.
22. Laemmli, U.K. Cleavage of structural proteins during the assembly of the head of bacteriophage T4. *Nature* **1970**, *227*, 680–685. [CrossRef] [PubMed]
23. Lowry, B.O.H.; Rosebrough, N.J.; Farr, A.L.; Randall, T.J. Protein measurement with Folin phenol reagent. *J. Biol. Chem.* **1951**, *193*, 265. [PubMed]
24. Sagu, S.T.; Nso, E.J.; Homann, T.; Kapseu, C.; Rawel, H.M. Extraction and purification of beta-amylase from stems of Abrus precatorius by three phase partitioning. *Food Chem.* **2015**, *183*, 144–153. [CrossRef] [PubMed]
25. He, L.; Park, S.-H.; Dang, N.D.H.; Duong, H.X.; Duong, T.P.C.; Tran, P.L.; Park, J.-T.; Ni, L. ; Park, K.-H. Characterization and thermal inactivation kinetics of highly thermostable ramie leaf β-amylase. *Enzyme Microb. Technol.* **2017**, *101*, 17–23. [CrossRef] [PubMed]

Sample Availability: Samples of SPA and Mal-mPEG5000-β-SPA are available from the authors.

© 2018 by the authors. Licensee MDPI, Basel, Switzerland. This article is an open access article distributed under the terms and conditions of the Creative Commons Attribution (CC BY) license (http://creativecommons.org/licenses/by/4.0/).

Article

Biodegradation of 7-Hydroxycoumarin in *Pseudomonas mandelii* 7HK4 via *ipso*-Hydroxylation of 3-(2,4-Dihydroxyphenyl)-propionic Acid

Arūnas Krikštaponis * and Rolandas Meškys

Department of Molecular Microbiology and Biotechnology, Institute of Biochemistry, Life Sciences Center, Vilnius University, Sauletekio al. 7, LT-10257 Vilnius, Lithuania; rolandas.meskys@bchi.vu.lt
* Correspondence: arunas.krikstaponis@bchi.vu.lt; Tel.: +370-616-17187

Received: 15 September 2018; Accepted: 10 October 2018; Published: 12 October 2018

Abstract: A gene cluster, denoted as *hcdABC*, required for the degradation of 3-(2,4-dihydroxyphenyl)-propionic acid has been cloned from 7-hydroxycoumarin-degrading *Pseudomonas mandelii* 7HK4 (DSM 107615), and sequenced. Bioinformatic analysis shows that the operon *hcdABC* encodes a flavin-binding hydroxylase (HcdA), an extradiol dioxygenase (HcdB), and a putative hydroxymuconic semialdehyde hydrolase (HcdC). The analysis of the recombinant HcdA activity in vitro confirms that this enzyme belongs to the group of *ipso*-hydroxylases. The activity of the proteins HcdB and HcdC has been analyzed by using recombinant *Escherichia coli* cells. Identification of intermediate metabolites allowed us to confirm the predicted enzyme functions and to reconstruct the catabolic pathway of 3-(2,4-dihydroxyphenyl)-propionic acid. HcdA catalyzes the conversion of 3-(2,4-dihydroxyphenyl)-propionic acid to 3-(2,3,5-trihydroxyphenyl)-propionic acid through an *ipso*-hydroxylation followed by an internal (1,2-C,C)-shift of the alkyl moiety. Then, in the presence of HcdB, a subsequent oxidative *meta*-cleavage of the aromatic ring occurs, resulting in the corresponding linear product (2E,4E)-2,4-dihydroxy-6-oxonona-2,4-dienedioic acid. Here, we describe a *Pseudomonas mandelii* strain 7HK4 capable of degrading 7-hydroxycoumarin via 3-(2,4-dihydroxyphenyl)-propionic acid pathway.

Keywords: 7-hydroxycoumarin; 3-(2,4-dihydroxyphenyl)-propionic acid; 3-(2,3,5-trihydroxyphenyl)-propionic acid; *ipso*-hydroxylase; *Pseudomonas mandelii*

1. Introduction

The compound 7-hydroxycoumarin (**5**), also known as umbelliferone, is one of the most abundant plant-derived secondary metabolites. It is a parent compound of other, more complex natural furanocoumarins and pyranocoumarins in higher plants [1–3]. In the case of plant damage, plants produce a high diversity of natural coumarins as a defense mechanism against insect herbivores as well as fungal and microbial pathogens [4]. For example, simple hydroxycoumarins have antibacterial activity against *Ralstonia solanacearum*, *Escherichia coli*, *Klebsiella pneumoniae*, *Staphylococcus aureus*, and *Pseudomonas aeruginosa* [4–8]. Despite the toxic effects of coumarins, it has been shown that microorganisms evolve to gain the ability to metabolize such compounds.

It has been shown previously that a number of soil microorganisms, such as *Pseudomonas*, *Arthrobacter*, *Aspergillus*, *Penicillium*, and *Fusarium* spp. can grow on coumarin (**1**) as a sole source of carbon [9–16]. The key intermediate during coumarin catabolism in bacteria is 3-(2-hydroxyphenyl)-propionic acid (**4**) [10,11,16]. The bioconversion of coumarin to 3-(2-hydroxyphenyl)-propionic acid can be achieved in two different metabolic pathways, as shown in Figure 1A. Bacteria belonging to the *Arthrobacter* genus enzymatically hydrolyze the lactone moiety to give 3-(2-hydroxyphenyl)-2-propenoic acid (**2**), and then reduce a double bond by using a NADH-dependent enzyme to yield 3-(2-hydroxyphenyl)-propionic acid [11]. In the case of *Pseudomonas* sp. 30-1 and *Aspergillus niger* ATCC 11394 cells, coumarin is initially reduced to dihydrocoumarin (**3**) by a NADH-dependent oxidoreductase and only then enzymatically hydrolyzed [9,10,14]. *Arthrobacter* and *Pseudomonas* species are capable of oxidizing 3-(2-hydroxyphenyl)-propionic acid to 3-(2,3-dihydroxyphenyl)-propionic acid by using specific flavin-binding aromatic hydroxylases [16]. However, no data are available on further conversions of 3-(2,3-dihydroxyphenyl)-propionic acid in these bacteria. Also, no microorganisms or enzymes implicated in the metabolism of any hydroxycoumarin have been identified to date.

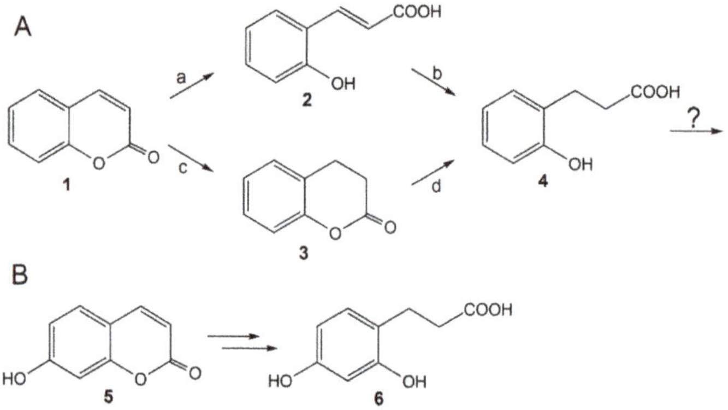

Figure 1. (**A**) Coumarin metabolic routes in *Arthrobacter* spp. (a, b), as well as *Pseudomonas* and *Aspergillus* spp. (c, d). (**B**) Proposed metabolic pathway of 7-hydroxycoumarin in *Pseudomonas* sp. 7HK4 bacteria. **1**—coumarin; **2**—3-(2-hydroxyphenyl)-2-propenoic acid; **3**—dihydrocoumarin; **4**—3-(2-hydroxyphenyl)-propionic acid; **5**—7-hydroxycoumarin; **6**—3-(2,4-dihydroxyphenyl)-propionic acid; a—putative coumarin hydrolase; b—NADH:o-coumarate oxidoreductase; c—dihydrocoumarin:NAD[NADP] oxidoreductase; d—putative dihydrocoumarin hydrolase.

Here we describe a *Pseudomonas* sp. 7HK4 strain capable of utilizing 7-hydroxycoumarin as the sole carbon and energy source, and a catabolic pathway of 7-hydroxycoumarin in these bacteria. Analysis of 7-hydroxycoumarin-inducible proteins led to the identification of the genome locus encoding 7-hydroxycoumarin catabolic proteins. The corresponding genes were cloned and heterologously expressed in *E. coli* system. The functions of the recombinant proteins were determined and enzyme activities towards various substrates were evaluated. We show that *Pseudomonas* sp. 7HK4 encodes a distinct 7-hydroxycoumarin metabolic route, which utilizes a flavin monooxygenase responsible for *ipso*-hydroxylation of 3-(2,4-dihydroxyphenyl)-propionic acid (**6**).

2. Results

2.1. Screening and Identification of 7-Hydroxycoumarin-Degrading Microorganisms

By the means of enrichment culture using various coumarin derivatives, an aerobic strain 7HK4 degrading 7-hydroxycoumarin was isolated from the garden soil in Lithuania (DSMZ accession number DSM 107615). This bacterium was tested for its ability to grow on several coumarin derivatives, such as coumarin, 3-hydroxycoumarin, 4-hydroxycoumarin, 6-hydroxycoumarin, 6-methylcoumarin, 6,7-dihydroxycoumarin, and 7-methylcoumarin, as the sole carbon and energy source in a minimal salt medium. However, of all the aforementioned compounds, the strain 7HK4 was able to utilize 7-hydroxycoumarin only. The strain utilized glucose, which was used as a control substrate in whole-cell reactions. Data on biochemical analysis of this strain are given in the Supplementary Material. The nucleotide sequence of 16S rRNA gene was determined by sequencing of the cloned DNA fragment, which was obtained by PCR amplification. The strain 7HK4 showed the highest

16S rDNA sequence similarity to that from *Pseudomonas* genus and was similar to 16S rDNA from *Pseudomonas mandelii* species according to the phylogenetic analysis, as shown in Figure S1 in the Supplementary Material.

2.2. Bioconversion Experiments by Using Whole Cells of Pseudomonas sp. 7HK4

Time course experiments using whole cells of *Pseudomonas* sp. 7HK4 pre-grown in the presence of 7-hydroxycoumarin showed that the cells converted coumarin, 6-hydroxycoumarin, and 6,7-dihydroxycoumarin in addition to 7-hydroxycoumarin, according to the changes in the UV-VIS spectra, as shown in Figure S2 in the Supplementary Material. UV absorption maxima were dropping down over time, although the biotransformation rates of 6-hydroxycoumarin, 6,7-dihydroxycoumarin, and coumarin were approximately 5-fold to 10-fold lower, respectively. After the completion of bioconversions, there were no visible spectra observed for the residual aromatic compound in the reaction mixtures with 7-hydroxycoumarin, except for the reactions with 6-hydroxycoumarin, 6,7-dihydroxycoumarin, and coumarin, which had non-disappearing UV absorption maxima at 260–270 nm wavelengths. These spectra are similar to that of 3-phenylpropionic acid (data not shown), suggesting that 7-hydroxycoumarin-induced 7HK4 strain can only catalyze the hydrolysis and reduction of lactone moiety of coumarin, 6-hydroxycoumarin, and 6,7-dihydroxycoumarin producing 3-(2-hydroxyphenyl)-propionic, 3-(2,5-dihydroxyphenyl)-propionic, and 3-(2,4,5-trihydroxyphenyl)-propionic acids, respectively, comparable to similar biotransformations in other microorganisms [9–16]. In addition, uninduced *Pseudomonas* sp. 7HK4 cells grown in the presence of glucose showed delayed and slower conversion of 7-hydroxycoumarin. The bioconversion process only started 0.5 h after the addition of substrate, suggesting that 7-hydroxycoumarin induced its own metabolism. In addition, uninduced cells did not catalyze any conversion or showed delayed and much slower biotransformations for coumarin, 6,7-dihydroxycoumarin, and 6-hydroxycoumarin, respectively. This demonstrates that in *Pseudomonas* sp. 7HK4 bacteria, the metabolism of 7-hydroxycoumarin is an inducible process.

It has been shown previously that 3-(2-hydroxyphenyl)-2-propenoic and 3-(2-hydroxyphenyl)-propionic acids, as shown in Figure 1B, are the intermediates in a known coumarin metabolic pathway in several microorganisms [9–11,14,16]. By analogy, it was suggested that 3-(2,4-dihydroxyphenyl)-propionic acid may be an intermediate metabolite in 7-hydroxycoumarin catabolism, as shown in Figure 1B. The cells of 7HK4 strain pre-cultivated in the presence of 7-hydroxycoumarin catalyzed no conversion of 3-(2-hydroxyphenyl)-2-propenoic or 3-(2-hydroxyphenyl)-propionic acid, however 3-(2,4-dihydroxyphenyl)-propionic acid was straightforwardly consumed. The HPLC-MS analysis of the bioconversion mixtures showed that *Pseudomonas* sp. 7HK4 cells cultivated in the presence of 7-hydroxycoumarin produce 3-(2,4-dihydroxyphenyl)-propionic acid as an intermediate metabolite, as shown in Figure 2. In 7HK4 bacteria grown on glucose, the aforementioned compound was not observed suggesting that 3-(2,4-dihydroxyphenyl)-propionic acid is an intermediate metabolite during 7-hydroxycoumarin degradation.

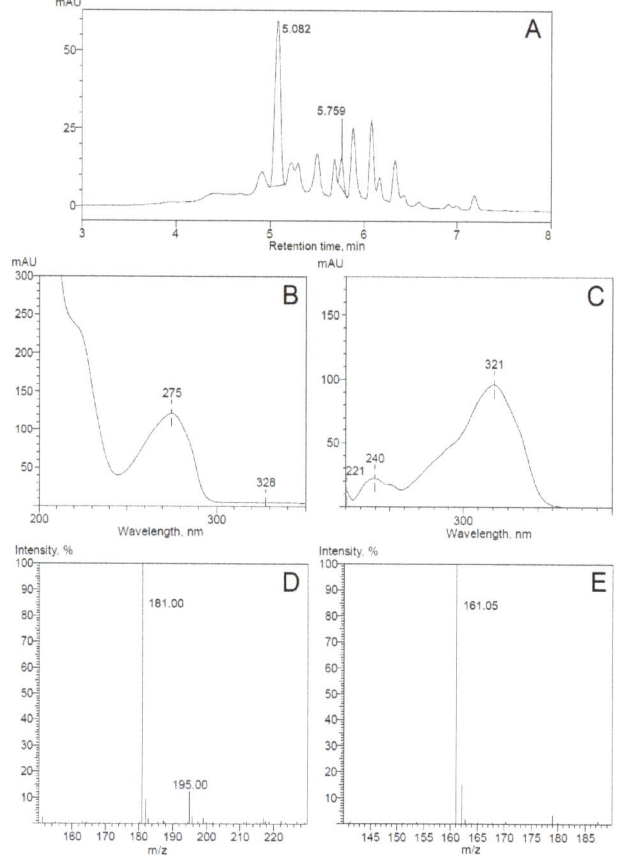

Figure 2. *Pseudomonas* sp. 7HK4 bacteria were grown in the presence of 7-hydroxycoumarin, and produced metabolites were analyzed by HPLC-MS. UV 254 nm trace of metabolites (**A**). UV and MS spectra of peaks with retention times 5.082 min (**B,D**) and 5.759 min (**C,E**). The negative ions [M−H]$^-$ generated are at m/z 181.00 (3-(2,4-dihydroxyphenyl)-propionic acid) and 161.05 (7-hydroxycoumarin).

2.3. Identification of 7-Hydroxycoumarin-Inducible Proteins

To elucidate which enzymes are involved in 7-hydroxycoumarin metabolism, *Pseudomonas* sp. 7HK4 cells were cultivated in a minimal medium supplemented with 7-hydroxycoumarin (0.3 mM) or glucose (0.3 mM) as the sole carbon and energy source. Several 7-hydroxycoumarin-inducible proteins of different molecular mass were observed by using SDS-PAGE analysis of cell-free extracts from *Pseudomonas* sp. 7HK4, as shown in Figure 3A.

Three bands corresponding to the inducible proteins of 23, 32, and 50 kDa, as shown in Figure 3A, were excised from the SDS-PAGE gel and analyzed by MS-MS sequencing. The genome sequence of strain 7HK4 was obtained via Illumina sequencing as described in Materials and Methods. For the identification of corresponding genes of inducible proteins, the sequences of the identified peptides were searched against the partial 7HK4 genome sequence. Thus, the genome fragment was discovered, encoding a 31.2 kDa protein containing 16 aa-long sequence, as shown in the bolded sequence in Figure 3A, identical to that found in 7-hydroxycoumarin-inducible ~32 kDa protein.

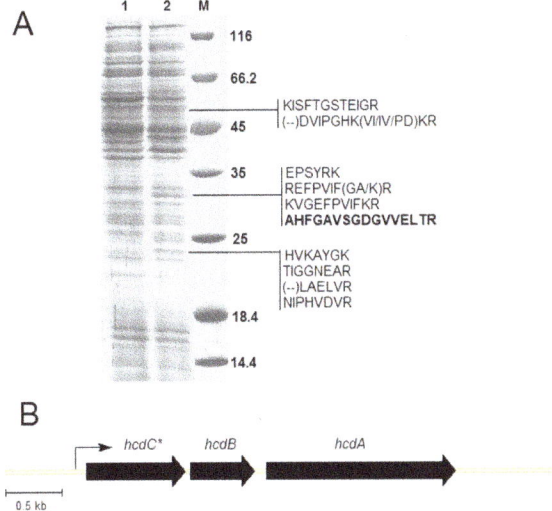

Figure 3. (**A**) SDS-PAGE analysis of cell-free extracts of *Pseudomonas* sp. 7HK4. Bacteria were cultivated in the presence of glucose (lane 1) or 7-hydroxycoumarin (lane 2). M—molecular mass ladder (kDa). The arrows indicate 7-hydroxycoumarin-inducible 23, 32, and 50 kDa proteins. The peptide sequences determined by MS-MS are given on the right. The sequence in bold was identified after the peptide sequences obtained by MS-MS were compared with the partially sequenced genome of *Pseudomonas* sp. 7HK4. (**B**) Organization of *hcd* genes in *Pseudomonas* sp. 7HK4 bacteria. The arrows indicate open reading frames (ORFs) encoding HcdA, HcdB, and HcdC. The gene for the 7-hydroxycoumarin-inducible protein of 31.2 kDa is marked by an asterisk.

2.4. Analysis of the Genome Locus Encoding the 7-Hydroxycoumarin-Inducible Protein

Analysis of *Pseudomonas* sp. 7HK4 genome sequences (35 contigs) showed that the inducible 31.2 kDa protein belongs to the fumarylacetoacetate (FAA) hydrolase family, which includes such enzymes as 2-keto-4-pentenoate hydratase, 2-oxohepta-3-ene-1,7-dioic acid hydratase, 2-hydroxy-6-oxo-6-phenylhexa-2,4-dienoate hydrolase, or bifunctional isomerases/decarboxylases (catechol pathway), as shown in Figure S3 in the Supplementary Material. The FAA family proteins are usually involved in the last stages of bacterial metabolism of aromatic compounds [17–19], suggesting that 31.2 kDa protein from *Pseudomonas* sp. 7HK4 participates in the final steps of 7-hydroxycoumarin metabolism, after oxidative cleavage of the aromatic ring.

Adjacent to the 31.2 kDa protein-encoding gene, two open reading frames (ORFs) were identified. All three genes are arranged on the same DNA strand, and are separated by short intergenic regions, suggesting that these genes are organized into an operon, as shown in Figure 3B. The putative operon was designated *hcdABC* (**h**ydroxy**c**oumarin **d**egrading operon), where *hcdC* encodes the 31.2 kDa protein. A BLAST analysis of *hcdA* and *hcdB* sequences revealed that these genes encode the putative FAD-binding hydroxylase and ring-cleavage dioxygenase, respectively. HcdA protein was not assigned to any family, but it showed similarity to a putative 2-polyprenyl-6-methoxyphenol hydroxylase, as shown in Figure S3 in the Supplementary Material. This type of enzyme belongs to class A of FAD-binding monooxygenases, which are involved in bacterial degradation of aromatic compounds [20–23]. The product of the hcdB gene belongs to the cl14632 superfamily that combines a variety of structurally related metalloproteins, including the type I extradiol dioxygenases, as shown in Figure S3 in the Supplementary Material. The type I extradiol dioxygenases catalyze the incorporation of both atoms of molecular oxygen into aromatic substrates that results in the cleavage of the aromatic rings [24,25].

2.5. Expression and Substrate Specificity of HcdA Hydroxylase

For further characterization of HcdA hydroxylase, the *hcdA* gene was amplified by PCR and cloned into the pET21b expression vector. The sequence was confirmed by Sanger sequencing. The recombinant C-terminally His$_6$-tagged protein was produced in *Escherichia coli* BL21, and purified by affinity chromatography. The purified enzyme migrated as a ~62 kDa band on SDS-PAGE, as shown in Figure S4A in the Supplementary Material, and had a bright yellow color with absorbance maxima at 380 and 450 nm wavelengths, as shown in Figure S4B in the Supplementary Material, suggesting that the protein contains a tightly bound flavin [26–28]. The gel-filtration showed that the purified HcdA protein is a monomer, as shown in Figure S4C in the Supplementary Material. The specificity for both flavin and nicotinamide cofactors was investigated. The HcdA hydroxylase was able to utilize either NADH or NADPH, although the oxidation rates of NADPH were almost 2-fold lower. Kinetic characterization of HcdA protein is presented in the Supplementary Material.

The activity of the HcdA enzyme was assayed in the presence of NADH cofactor against various substrates. The highest rate of oxidation of NADH was recorded in the presence of 3-(2,4-dihydroxyphenyl)-propionic acid. A 40-fold lower rate was observed when *trans*-2,4-dihydroxycinnamic acid was used as the substrate. The HcdA was not active towards *trans*-cinnamic, *cis*-2,4-dihydroxycinnamic, 3-(2-hydroxyphenyl)-propionic, 3-(2-hydroxyphenyl)-2-propenoic, 3-(4-hydroxyphenyl)-2-propenoic, 3-(3-hydroxyphenyl)-2-propenoic, 3-(2-bromophenyl)-propionic, 3-(2-nitrophenyl)-propionic, and 3-phenylpropionic acids, cinnamyl alcohol, pyrocatechol, 3-methylcatechol, 4-methylcatechol, 2-propylphenol, 2-propenylphenol, 2-ethylphenol, o-cresol, o-tyrosine, resorcinol, 2,3-dihydroxypyridine, 2-hydroxy-4-aminopyridine, N-methyl-2-pyridone, N-ethyl-2-pyridone, N-propyl-2-pyridone, N-butyl-2-pyridone, indoline, and indole. These data demonstrate that the HcdA is a highly specific monooxygenase, which shows a strong preference to 3-(2,4-dihydroxyphenyl)-propionic acid. The addition of His$_6$-tag did not affect the enzymatic activity of the HcdA protein.

The product of the reaction catalyzed by the HcdA hydroxylase was analyzed by UV-VIS absorption spectroscopy and HPLC-MS. A new UV absorption maximum was observed at 340 and 490 nm upon addition of 3-(2,4-dihydroxyphenyl)-propionic acid to the reaction mixture. The consequent red coloring was observed, which indicated the presence of *para*- or *ortho*-quinone [29], a presumed product of the autoxidation of the corresponding hydroquinone. The same coloration was also observed in vivo when *Pseudomonas* sp. 7HK4 cells were grown in an excess of 7-hydroxycoumarin and when *Escherichia coli* BL21 cells harboring the p4pmPmo plasmid were cultured in the presence of 3-(2,4-dihydroxyphenyl)-propionic acid. HPLC-MS analysis of the reaction mixtures of both in vitro and in vivo bioconversions confirmed the formation of 3-(trihydroxyphenyl)-propionic acid and quinone, found [M−H]$^-$ masses were 197.00 (traces seen only in vivo) and 195.00, respectively, as shown in Figure 4. However, the structure of the product was not confirmed at this stage by chemical analysis since it was not possible to chromatographically separate the product from the reaction mixture. The substrate and product have similar structures and chemical properties; therefore, both were detected under the same HPLC-MS chromatogram peak with retention time of ~5.5 min.

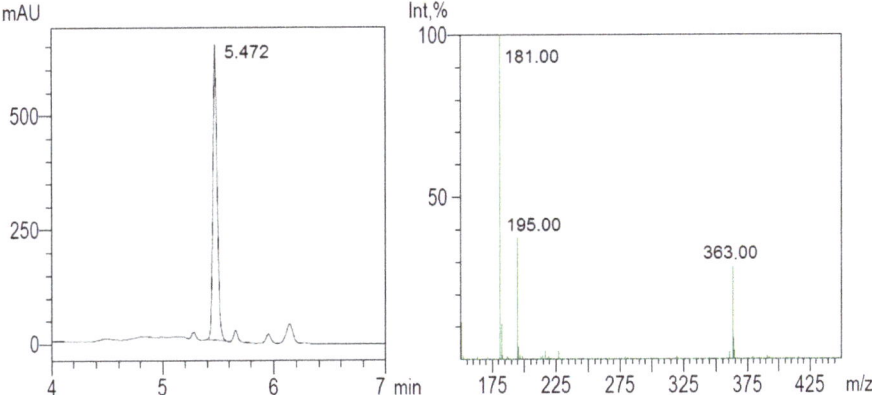

Figure 4. HPLC-MS analysis of 3-(2,4-dihydroxyphenyl)-propionic acid bioconversion mixture in vitro. UV 254 nm trace of 3-(2,4-dihydroxyphenyl)-propionic acid and its hydroxylated product under the same peak with retention time 5.472 min (on the left), and MS spectrum of the dominant peak (on the right). The negative ions [M−H]⁻ generated are at m/z 181 (3-(2,4-dihydroxyphenyl)-propionic acid), 195 (product of 3-(trihydroxyphenyl)-propionic acid autooxidation), and 363 (dimer of 3-(2,4-dihydroxyphenyl)-propionic acid).

2.6. Expression and Substrate Specificity of HcdB Dioxygenase

HcdB dioxygenase was expressed from the plasmid pTHPPDO in *E. coli* BL21. All attempts to purify HcdB aerobically resulted in the loss of the enzymatic activity, even in the presence of organic solvents, such as glycerol, ethanol, and acetone that were known as stabilizers for similar enzymes [30–32]. The addition of dithiothreitol and ferrous sulfate to the aerobically purified protein did not restore its activity, although it had been shown to activate and/or stabilize other extradiol dioxygenases [30–34].

Therefore, due to the highly unstable nature of the HcdB enzyme, all the activity measurements were carried out in vivo using the whole cells of *E. coli* BL21 transformed with the pTHPPDO plasmid. The bioconversion of pyrocatechol, 3-methylcatechol, 3-methoxycatechol, 4-methylcatechol, and caffeic acids in each case produced yellow products with the absorbance maxima expected for the proximal *meta*-cleavage products of these catechols, as shown by the solid line in Figure 5 [35–37]. A weak yellow color was visible in bioconversion mixture containing 3-(2,3-dihydroxyphenyl)-propionic acid, and no changes in color took place, but the absorbance maxima shifts were observed during reactions with pyrogallol, gallacetophenone, 2′,3′-dihydroxy-4′-methoxyaceto-phenone hydrate, 3,4-dihydroxybenzoic acid, 2,3,4-trihydroxybenzoic acid, and 2,3,4-trihydroxy-benzophenone, as shown in Figure S11 in the Supplementary Material. No activity was observed with 1,2,4-benzenetriol and 6,7-dihydroxycoumarin. The *E. coli* cells without the *hcdB* gene showed no activity towards the aforementioned compounds. The expected shift of peaks of the UV-VIS spectra of the reaction products to a shorter wavelength were observed after acidification of the reaction mixtures containing pyrocatechol, 3-methylcatechol, 3-methoxycatechol, 4-methylcatechol, and caffeic acid, as shown by the dashed line in Figure 5 [38,39]. These findings revealed that *hcdB* encodes an extradiol dioxygenase, which can utilize a number of differently substituted catechols.

Furthermore, the HcdB dioxygenase was co-expressed with the HcdA hydroxylase in *E. coli* BL21 cells. The activity of those cells towards 3-(2,4-dihydroxyphenyl)-propionic acid was analyzed. No coloration occurred during the incubation of over 72 h, compared to the formation of a reddish bioconversion product by *E. coli* cells containing the *hcdA* gene only. Products of 3-(2,4-dihydroxyphenyl) propionic acid conversion by HcdA and HcdB were analyzed by HPLC-MS. Ions with masses 181.00, 195.00, 197.00 ([M−H]⁻) were not detected showing a complete conversion of substrate and its hydroxylated forms. Also, none of the expected ions ([M−H]⁻ 229.00 or [M+H]⁺ 231.00) were observed for a presumed product of the oxidative cleavage of 3-(trihydroxyphenyl)-propionic acid.

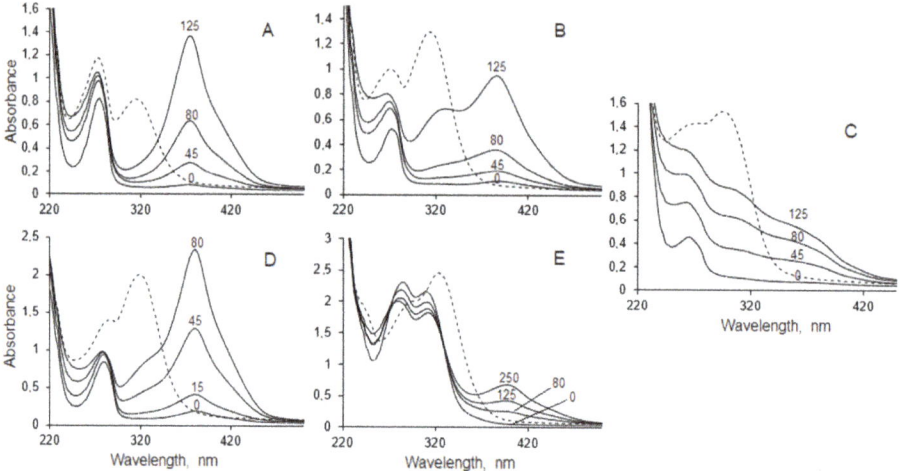

Figure 5. Biotransformations of pyrocatechol (**A**), 3-methylcatechol (**B**), 3-methoxycatechol (**C**), 4-methylcatechol (**D**), and caffeic acid (**E**) by whole cells of *E. coli* BL21 containing *hcdB* gene. Biotransformations were carried out in 50 mM potassium phosphate buffer pH 7.5 (solid line) at 30 °C with 0.5–1 mM of substrate. Incubation time is shown in min. Dashed lines indicate peak shifts to a shorter wavelength after acidification of reaction mixtures.

2.7. Isolation and Identification of 3-(2,4-Dihydroxyphenyl)-propionic Acid Bioconversion Product

Due to difficulties in detecting colorless *meta*-cleavage product of catechol derivative, and since no reasonable mass spectra could be registered, we decided to transform a cleavage product into the derivative of picolinic acid by incubation with NH₄Cl as described in Materials and Methods. The formation of a picolinic acid derivative was proven by HPLC-MS analysis, which showed the formation of [M−H]⁻ ion 210.00 mass, as shown in Figure 6, corresponding to the addition of NH₃ to the *meta*-cleavage product and the loss of two H₂O molecules [40].

The ^{1}H NMR spectrum of this derivative [δ 7.73 (s, 1H), 6.89 (s, 1H), 2.65 (t, J = 7.4 Hz, 2H), 2.38 (t, J = 7.4 Hz, 2H)] showed a set of two aryl protons that, from the coupling pattern (singlet + singlet), were in *meta*- or *para*-positions to each other on the aromatic ring [41], as shown in Figure S12 in the Supplementary Material. The appearance in the spectrum of two triplets with chemical shifts of 2.38 and 2.65 ppm indicated the presence of four methylene protons [41]. The ^{13}C NMR spectrum [δ 181.51, 179.99, 171.01, 143.14, 136.74, 130.11, 115.22, 35.50, 24.00] showed two sp³ carbons with chemical shifts of 24.00 and 35.50 ppm, and three sp² carbons of carbonyl groups with chemical shifts of 171.01, 179.99, and 181.51 ppm, respectively, as shown in Figure S13 in the Supplementary Material. The carbonyl carbon atoms were the most strongly deshielded and their resonances formed a separate region at the highest frequency. Another four sp² carbon signals were in the aromatic carbon region [41,42]. The presence of the third carbonyl group indicated the formation of *oxo*-pyridine, for which six possible theoretical structures of *oxo*-picolinic acid derivative were presumed, as shown in Figure 7. Since the ^{1}H NMR spectrum showed a set of two singlet aryl protons in *meta*- or *para*-positions to each other, only structures **7** and **9**, as shown in Figure 7, were further analyzed. Besides, pyridine aromatic carbons are usually differentiated into two resonances at higher field (C-3/5, *meta* position) and three at lower field (C-2/6, *ortho* position; C-4, *para* position), where the electron-withdrawing effect of nitrogen is effective [41,42]. The chemical shift of 115.22 ppm showed that the analyzed compound had relatively strongly shielded unsubstituted aromatic carbon, which should be in *meta*-position from nitrogen, in *ortho*-position from the carbonyl group, and in *meta*- or *para*-position from the carboxyl group (C-5 atom), as shown in Figure S14 in the Supplementary Material [43,44]. This led to the conclusion that structure **9**, as shown in Figure 7, 6-(2-carboxyethyl)-4-oxo-1,4-dihydropyridine-2-carboxylic acid, was formed during incubation of *meta*-cleavage product of catechol derivative with NH_4Cl, as depicted in Scheme 1. These data allowed the reconstruction of the consecutive oxidation of 3-(2,4-dihydroxyphenyl)-propionic acid catalyzed by the HcdA and HcdB enzymes. Hence, 3-(2,3,5-trihydroxyphenyl)-propionic acid (**14**) was the product of oxidation of 3-(2,4-dihydroxyphenyl)-propionic acid by the HcdA hydroxylase. The molecular mass of 198.17 of 3-(2,3,5-trihydroxyphenyl)-propionic acid and capability to form *para*-quinone agreed with the UV-VIS and HPLC-MS data on the bioconversion of 3-(2,4-dihydroxyphenyl)-propionic by the HcdA hydroxylase. The formation of 3-(2,3,5-trihydroxyphenyl)-propionic acid from 3-(2,4-dihydroxyphenyl)-propionic acid was possible only through oxidative *ipso*-rearrangement, a unique reaction where *ipso*-hydroxylation (**13**) of the 3-(2,4-dihydroxyphenyl)-propionic acid takes place with a simultaneous shift of the propionic acid group to the vicinal position, as shown in Scheme 1 [45–48]. During the second step, 3-(2,3,5-trihydroxyphenyl)-propionic acid was cleaved by HcdB extradiol dioxygenase at the *meta*-position leading to the formation of (2*E*,4*E*)-2,4-dihydroxy-6-oxonona-2,4-dienedioic acid (**15**). The further imine formation and tautomerization in the presence of ammonium ions [38,49] led to 6-(2-carboxyethyl)-4-oxo-1,4-dihydropyridine-2-carboxylic acid (**19**), as shown in Scheme 1.

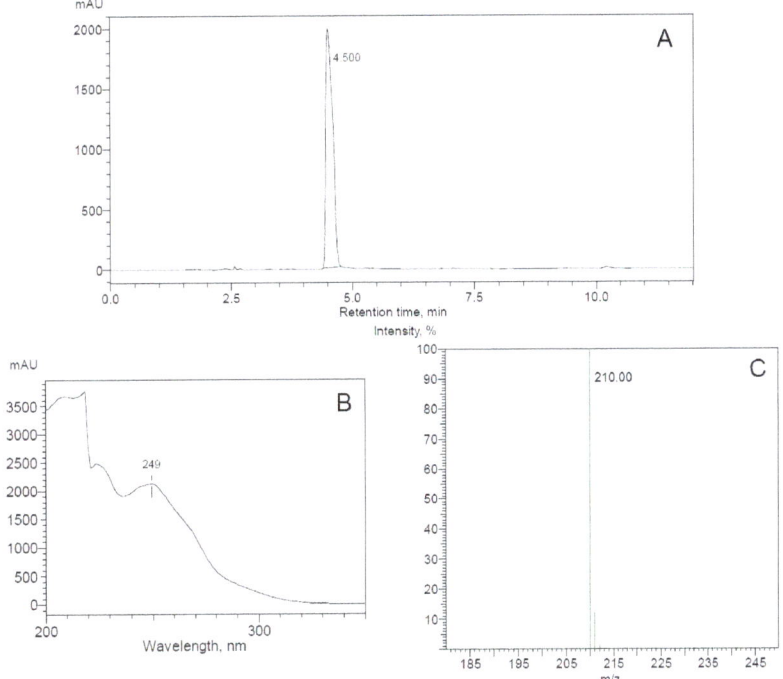

Figure 6. HPLC-MS analysis of 3-(2,4-dihydroxyphenyl)-propionic acid bioconversion mixture in vivo. UV 254 nm trace of picolinic acid derivative with retention time of 4.500 min (**A**), UV spectrum (**B**) and MS spectrum (**C**) of the dominant peak. The negative ions [M−H]⁻ generated are at *m/z* 210.00.

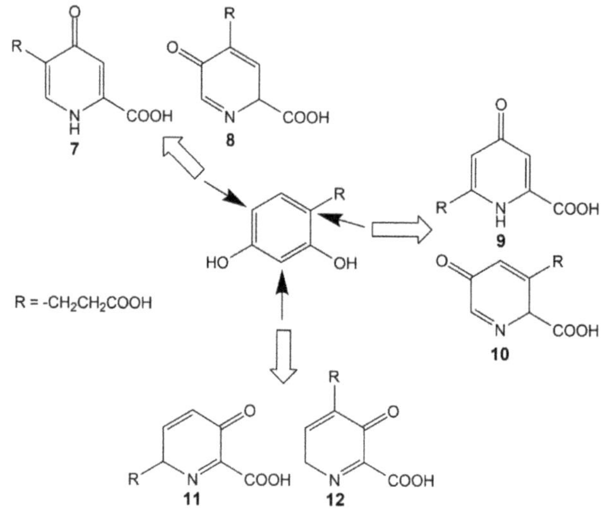

Figure 7. Suggested structures of *oxo*-picolinic acid derivative, formed during oxidative ring cleavage of 3-(2,4-dihydroxyphenyl)-propionic acid and conversion of the ring fission product. Solid lines indicate possible positions of hydroxylation by HcdA enzyme; hollow arrays indicate the probable *oxo*-picolinic acid derivatives forming after hydroxylation at each position.

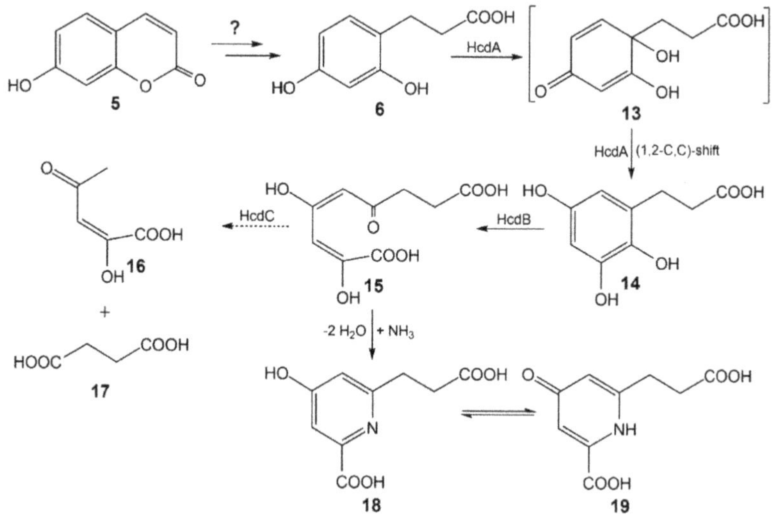

Scheme 1. The proposed metabolic pathway of 7-hydroxycoumarin in *Pseudomonas* sp. 7HK4 cells. Incubation of the compound **5** with NH$_4$Cl gives picolinic acid derivative. **5**—7-hydroxycoumarin; **6**—3-(2,4-dihydroxyphenyl)-propionic acid; **13**—3-(1,2-dihydroxy-4-oxocyclohexa-2,5-dienyl)-propanoic acid; **14**—3-(2,3,5-trihydroxyphenyl)-propionic acid; **15**—(2E,4E)-2,4-dihydroxy-6-oxonona-2,4-dienedioic acid; **16**—(E)-2-hydroxy-4-oxopent-2-enoic acid; **17**—succinic acid; **18**—6-(2-carboxyethyl)-4-hydroxypicolinic acid; **19**—6-(2-carboxyethyl)-4-oxo-1,4-dihydropyridine-2-carboxylic acid; HcdA—3-(2,4-dihydroxyphenyl)-propionic acid 1-monooxygenase; HcdB—3-(2,3,5-trihydroxyphenyl)-propionic acid 1,2-dioxygenase; HcdC—putative (2E,4E)-2,4-dihydroxy-6-oxonona-2,4-dienedioic acid hydrolase. The dashed arrow indicates a hypothetical reaction.

2.8. Expression of the HcdC Protein

HcdC was expressed from the plasmid p2K4PH in *E. coli* BL21. The addition of *E. coli* extracts containing the HcdC protein did not cause decolorization of the *meta*-cleavage products of 3-(2,3-dihydroxyphenyl)-propionic acid, pyrocatechol, 3-methylcatechol, or 4-methylcatechol formed in the presence of the HcdB extradiol dioxygenase. Also, the addition of NAD(P)$^+$ to these reaction mixtures did not induce decolorization of the *meta*-cleavage products [50]. To confirm the function of HcdC, *E. coli* BL21 cells were transformed with p4pmPmo and pCDF-BC plasmids. The expression of *hcdA*, *hcdB*, and *hcdC* genes in *E. coli* cells was confirmed by SDS-PAGE, the enzymes migrated as 62, 31, and 20 kDa bands, respectively, as shown in Figure S15 in the Supplementary Material.

The bioconversion of 3-(2,4-dihydroxyphenyl)-propionic acid was conducted in *E. coli* cells containing all three recombinant proteins. Later, the reaction mixture was incubated with NH$_4$Cl, and the reaction products were analyzed by HPLC-MS. The ions [M−H]$^-$ and [M+H]$^+$ with masses of 210.00 and 212.00, respectively, were not detected, compared to the bioconversion mixture with *E. coli* cells containing *hcdAB* genes only. This showed a complete conversion of *meta*-cleavage product of the catechol derivative, therefore no picolinic acid derivative could be obtained. No reaction products of (2E,4E)-2,4-dihydroxy-6-oxonona-2,4-dienedioic acid hydrolysis by the HcdC protein were identified. We suggest that the later compound was converted to succinic acid (**17**), which entered the Krebs cycle, and (E)-2-hydroxy-4-oxopent-2-enoic acid (**16**), which could be further converted by *E. coli* cells, thus complicating the extraction of these reaction products. Nevertheless, it may be concluded that all three enzymes encoded by the *hcdABC* operon are responsible for the catabolism of 3-(2,4-dihydroxyphenyl)-propionic acid in *Pseudomonas* sp. 7HK4 bacteria.

3. Discussion

Although coumarins are widely abundant in nature and are intensively used in biotechnology as precursor compounds [51], the metabolic pathways in microorganisms are still not known in sufficient detail. In this study, *Pseudomonas* sp. 7HK4 strain was isolated from soil and it was shown that these bacteria can utilize only 7-hydroxycoumarin as a sole source of carbon and energy. Several other coumarin derivatives were also tested, but none of those substrates support the growth of *Pseudomonas* sp. 7HK4. However, the experiments with the 7-hydroxycoumarin-induced whole cells show that *Pseudomonas* sp. 7HK4 has the enzymes that are able to transform coumarin and 6-hydroxycoumarin. The products of these biotransformations give UV spectra similar to the UV spectrum of 3-phenylpropionic acid, suggesting that *Pseudomonas* sp. 7HK4 bacteria can catalyze hydrolysis and reduction of lactone moiety of these substrates. Furthermore, 3-(2,4-dihydroxyphenyl)-propionic acid has been identified as an intermediate metabolite during biotransformations of 7-hydroxycoumarin. This finding agrees with the data published for the conversion of coumarin by *Pseudomonas* spp., *Arthrobacter* spp., and *Aspergillus* spp., which all catalyze the hydrolysis and reduction of coumarin producing 3-(2-hydroxyphenyl)-propionic acid as the main intermediate [9–11,14]. However, compared with *P. mandelii* 7HK described in this study, little is known about the bioconversion of other coumarin derivatives as well as further conversions of 3-(2-hydroxyphenyl)-propionic acid in bacteria listed above.

The analysis of the 7-hydroxycoumarin-inducible proteins lead to the identification of the genomic locus *hcdABC* encoding the enzymes required for 3-(2,4-dihydroxyphenyl)-propionic acid degradation. A BLAST search uncovered that *hcdA*, *hcdB*, and *hcdC* encode an aromatic flavin-binding hydroxylase, a ring-cleavage dioxygenase and a hydrolase, respectively. Biochemical analysis of the HcdA protein confirmed its similarity to class A FAD-binding enzymes (FMOs) [20–23,26–28]. HcdA is functionally related to previously well-described hydroxylases, such as OhpB monooxygenase from *Rhodococcus* sp. V49 [52], MhpA monooxygenase from *E. coli* K-12 [53] and *Comamonas testosteroni* TA441 [54], HppA monooxygenase from *Rhodococcus globerulus* PWD1 [31], and also *para*-hydroxybenzoate hydroxylase (PHBH) from *P. fluorescens* [55]. HcdA appears to have a high specificity for 3-(2,4-dihydroxyphenyl)-propionic acid, converting it to 3-(2,3,5-trihydroxyphenyl)-propionic acid, and a 40-fold lower activity towards *trans*-2,4-dihydroxycinnamic acid. Other structurally related substrates are not used by HcdA. Unlike HcdA, other related FMOs have a broader specificity for substrates. For example, OhpB monooxygenase is capable to oxidize 2-hydroxy-, 3-hydroxyphenylpropionic and cinnamic acids [52]. The HppA enzyme is more specific to 3-hydroxyphenylpropionic, but 4-chlorophenoxyacetic as well as 4-methyl-2-chlorophenoxyacetic acids are also oxidized [31]. On the other hand, all described FMOs including HcdA are NAD(P)H dependent, which reduces flavin for the hydroxylation of substrates [20–23]. The narrow specificity of HcdA to its natural substrate is typical for class A flavoproteins and shows the importance of HcdA enzyme in metabolism of 7-hydroxycoumarin. The kinetic analysis of HcdA yields K_m values of 50.10 ± 3.50 µM and 13.00 ± 1.20 µM for NADH and 3-(2,4-dihydroxyphenyl)-propionic acid, respectively. However, no kinetic parameters have been reported for OhpB, MhpA, and HppA hydroxylases for comparison, though PHBH has been shown to have a k_{cat} of 22.83 s^{-1} [56], which is 3-fold higher than a turnover number for the HcdA enzyme.

The hydroxylated product of the HcdA protein was analyzed by oxidizing it with the HcdB dioxygenase, followed by a chemical modification to the corresponding derivative of picolinic acid. The structure of the later compound was confirmed by ^{1}H NMR and ^{13}C NMR spectra. This allowed the reconstruction of the reaction products of both HcdA and HcdB enzymes. It was shown that a hydroxylation of 3-(2,4-dihydroxyphenyl)-propionic acid occurs at *ipso*-position of phenolic ring followed by internal rearrangements involving (1,2-C,C)-shift (NIH shift) of propionic acid moiety, hence forming 3-(2,3,5-trihydroxyphenyl)-propionic acid, as shown in Scheme 1. This would explain both the high specificity of HcdA enzyme for substrates with *para*-substituted phenol and inability of *Pseudomonas* sp. 7HK4 bacteria to utilize coumarin derivatives other than 7-hydroxycoumarin as the sole source of carbon and energy. Only a few classes of enzymes are able to catalyze *ipso*-reactions: laccases, peroxidases, dioxygenases, glutathione S-transferases (GST), cytochrome P450-dependent monooxygenases (CYP), and flavin-dependent monooxygenases. Among the known examples of *ipso*-enzymes, there are dioxygenases from *Comamonas testosteroni* T2 and *Sphingomonas* sp. strain RW1, which are involved in the desulfonation of 4-sulfobenzoate by *ipso*-substitution [57,58]. A rat liver CYP system is able to convert *p*-chloro, *p*-bromo, *p*-nitro, *p*-cyano, *p*-hydroxymethyl, *p*-formyl, and *p*-acetyl phenols to hydroquinone by *ipso*-hydroxylation [59]. GST is capable of catalyzing desulfonylation of sulfonylfuropyridine compounds by nucleophilic attack of the glutathione sulfur atom at *ipso*-position [60]. These are the examples of electrophilic or nucleophilic *ipso*-substitution reactions, however in some cases, a primary *ipso*-group is not eliminated, and instead it is shifted to *meta*-position. NIH shift restabilizes cyclohexadienone intermediate, because it leads to a rearomatization [45–48]. We showed that, similarly to flavin-dependent monooxygenases from *Sphingomonas* sp. TTNP3 and *Sphingobium xenophagum* strains, which are responsible for the degradation of alkylphenols, such as bisphenol A, octylphenol, *t*-butylphenol, *n*-octyloxyphenol, and *t*-butoxyphenol, HcdA-catalyzed reaction involves a NIH shift. Usually, NIH shift products are formed during the side reactions, and these internal rearrangements of an alkyl group upon the *ipso*-hydroxylation are spontaneous and non-enzymatic in *Sphingomonas* sp. strains [47,61]. An interesting novelty is that HcdA hydroxylase produces only one product, which has the *ipso*-group shifted to the *meta*-position. Therefore, we propose that in the case of HcdA, NIH shift occurs enzymatically, but not spontaneously or by a dienone-phenol rearrangement mechanism [62], since all bioconversions were performed under neutral or basic conditions. Although a further investigation is needed to determine the exact mechanism of the HcdA enzyme activity.

The analysis of the extradiol dioxygenase HcdB shows that this enzyme has a lower specificity for substrates. It catalyzes a conversion of the hydroxylated product of the HcdA enzyme to (2E,4E)-2,4-dihydroxy-6-oxonona-2,4-dienedioic acid, as shown in Scheme 1. Also, HcdB is capable oxidizing pyrocatechol, 3-methylcatechol, 3-methoxycatechol, 4-methylcatechol, 3-(2,3-dihydroxyphenyl)-propionic, and caffeic acids using a *meta*-cleavage mechanism forming yellow products. HcdB belongs to type I, class II extradiol dioxygenases [24,25], and is functionally related to OhpD catechol 2,3-dioxygenase from *Rhodococcus* sp. V49 [52], MhpB extradiol dioxygenase from *E. coli* K-12 [53,63], MpcI extradiol dioxygenase from *Alcaligenes eutrophus* [63], HppB extradiol dioxygenase from

Rhodococcus globerulus PWD1 [31], and DbfB 2,2′,3-trihydroxybiphenyl dioxygenase from *Sphingomonas* sp. RW1 [30].

Finally, we demonstrate that (2*E*,4*E*)-2,4-dihydroxy-6-oxonona-2,4-dienedioic acid, the ring cleavage product of HcdB protein, is subsequently hydrolyzed by the putative HcdC hydroxymuconic semialdehyde hydrolase. This enzyme has a low sequence homology to any of the previously characterized enzymes from *Rhodococcus* sp. V49 [52], *E. coli* K-12 [53], *Comamonas testosteroni* TA441 [55], or *Rhodococcus globerulus* PWD1 [31], therefore further investigation is needed to elucidate the exact mechanism of the hydrolysis and the specificity of the HcdC enzyme for substrates.

In summary, here we report a 7-hydroxycoumarin catabolic pathway in *Pseudomonas* sp. 7HK4 bacteria. New metabolites and genes responsible for the degradation of 3-(2,4-dihydroxyphenyl)-propionic acid have been isolated and identified. Our results show that the degradation of 7-hydroxycoumarin in *Pseudomonas* sp. 7HK4 involves a distinct metabolic pathway, compared to the previously characterized coumarin catabolic routes in *Pseudomonas*, *Arthrobacter*, and *Aspergillus* species [9–16]. It has been shown that *Pseudomonas* sp. 7HK4 bacteria employ unique flavin-binding *ipso*-hydroxylase for the oxidation of the aromatic ring of 3-(2,4-dihydroxyphenyl)-propionic acid. None of the proteins described in this paper have substantial sequence homology to the previously characterized enzymes implicated in the degradation of structurally similar substrates, such as 3-(2-hydroxyphenyl)-propionic acid in *Rhodococcus* sp. V49 [52], 3-(3-hydroxyphenyl)-propionic acid and 3-hydroxycinnamic acid in *E. coli* K-12 [53], *Comamonas testosteroni* TA441 [54], *Rhodococcus globerulus* PWD1 [31], or even 4-hydroxyphenylacetate in *Escherichia coli* W [64]. Thus, our results provide a fundamentally new insight into the degradation of hydroxycoumarins by the soil microorganisms. In addition, the discovered new bacteria and enzymes can be further employed for the development of novel biocatalytic processes useful for industry.

4. Materials and Methods

4.1. Bacterial Strains, Plasmids, and Reagents

Pseudomonas sp. 7HK4 bacterial strain capable of using 7-hydroxycoumarin as the sole source of carbon and energy was isolated from soil by enrichment in mineral medium containing 0.05% of 7-hydroxycoumarin. For cloning purposes, *E. coli* DH5α bacteria (φ80 *lacZ*ΔM15 Δ(*lacZY*-*argF*)U169 *deoR recA1 endA1 hsdR17*(r_K-m_K+) *supE44 thi-1 gyrA96 relA1*) (Thermo Fischer Scientific, Lithuania) were used. *E. coli* BL21 (DE3) bacteria (F- *ompT gal dcm lon hsdS*$_B$(r_B-m_B-) λ(DE3) [*lacI lacUV5*-T7 gene 1, *ind1*, *sam7*, *nin5*]) (Novagen, Darmstadt, Germany) were used for gene expression studies.

All reagents used during this study are listed in Table S1 and plasmids are described in Table S2 in the Supplementary Material.

4.2. Bioconversions with Whole Cells

Pseudomonas sp. 7HK4 bacteria were grown in mineral medium supplemented with 0.05% (*w*/*v*) of 7-hydroxycoumarin or glucose, as the sole carbon and energy source, at 30 °C with rotary aeration (180 rpm) for 48 h. *E. coli* BL21 (DE3) bacteria containing recombinant genes were grown in 30 mL of Brain Heart Infusion (BHI) medium at 30 °C and 180 rpm overnight. High density bacterial culture was centrifuged and resuspended in 30 mL of minimal C-750501 medium, in which the synthesis of proteins was induced with 1 mM of isopropyl-β-D-1-thiogalactopyranoside (IPTG) after 1.5 h of incubation at 20 °C and 180 rpm [65]. Incubation at 20 °C was continued for another 24 h. Both *Pseudomonas* sp. 7HK4 and *E. coli* cells were sedimented by centrifugation (3220× *g*, 15 min). The collected cells were washed twice with 15 mL of 0.9% NaCl solution. For whole-cell conversion experiments, cells from 20 mL of culture were resuspended in 1 mL of 50 mM potassium phosphate buffer (pH 7.2). All small-scale bioconversions with whole cells were made in 50 mM potassium phosphate buffer, pH 7.2, which contained 1–2 mM of the substrate. The reaction mixtures were kept in a thermoblock at

30 °C and 500 rpm. Bioconversion mixtures were centrifuged for 2 min at 10,000× g, and 100 µL of the supernatant were analyzed by UV-VIS spectroscopy (range 200–600 nm). Measurements were repeated to record the changes in the absorption intensity over time. All measurements were performed with PowerWave XS microplate reader.

4.3. Preparation of cell-free extracts

Cells were sedimented by centrifugation (3220× g, 15 min). The biomass was resuspended in 3 mL of 50 mM potassium phosphate buffer (pH 7.2). The cells were disrupted by pulse-mode sonication (3 min duration and 1 s cycles) at 4 °C. Cell debris was removed by centrifugation (4 °C, 16,100× g, 15 min).

4.4. Protein Purification

Proteins were purified with Äkta purifier 900 chromatography systems (GE Healthcare, Helsinki, Finland). Cell-free extracts were loaded onto a Ni^{2+} Chelating HiTrapTM HP column (1–5 mL) (GE Healthcare, Finland) equilibrated with 50 mM potassium phosphate buffer, pH 7.2, at 1.0 mL/min. The column was washed with at least 3 volumes of the same buffer. Then the bound proteins were eluted with 0.5 M imidazole in 50 mM potassium phosphate buffer, pH 7.2. The fractions containing the purified enzyme were combined and dialyzed against the 50 mM potassium phosphate buffer, pH 7.2, at 4 °C overnight. Proteins were analyzed by SDS-PAGE according to Laemmli [66]. Protein concentration was determined by the Lowry method [67].

4.5. Preparation of Proteins from a Polyacrylamide Gel for Mass Spectrometric Analysis

Proteins were fractionated on a SDS-polyacrylamide gel. After Coomassie blue R-250 staining, protein samples were extracted from the gel as described in Reference [68] with minor changes. Protein bands were excised from the gel with a razor, and the gel was then destained twice with 200 µL of 25 mM ammonium bicarbonate and 50% acetonitrile solution for 30 min at 37 °C. Protein disulfide bonds were reduced with 40 µL of 10 mM dithiothreitol (DTT) for 45 min 60 °C, followed by incubation with 30 µL of 100 mM iodoacetamide for 1 h at room temperature in the dark to alkylate free cysteines. Gel slices were washed again twice with 100 µL 25 mM ammonium bicarbonate and 50% acetonitrile solution for 15 min at 37 °C, dehydrated by adding 50 µL 100% acetonitrile and dried using a vacuum centrifuge. Gel pieces were incubated with up to 40 µL of activated trypsin (10 ng/µL) at 37 °C overnight. The next day, the supernatant was saved and the peptides were extracted from the gel by incubating gel slices in two consecutive changes of 50 µL of 5% trifluoroacetic acid and 50% acetonitrile solution for 1 h at 37 °C. Combined supernatants were dried using a vacuum centrifuge at 30 °C. Lyophilized peptides were dissolved in 20 µL of 0.1% trifluoroacetic acid solution. Peptides purified and concentrated using Millipore C18 ZipTips.

4.6. Enzyme Assay

Activity of 3-(2,4-dihydroxyphenyl)-propionic acid hydroxylase was measured spectrophotometrically by monitoring absorption changes of the reaction mixture at 340 nm wavelength due to the oxidation of either NADH or NADPH ($\varepsilon 340$ = 6220 M^{-1} cm^{-1}) after the addition of the substrate. The activity measurements were made with cell-free extracts or the purified protein. All measurements of the enzyme activity were carried out at room temperature in 1 mL of reaction mixture, containing 25–50 mM tricine or potassium phosphate buffers (pH 7.8), 100 µM NADH or NADPH, and 150 µM aromatic substrate. One unit of activity was defined as the amount of the enzyme that catalyzed the oxidation of 1 µmol of NADH or NADPH per minute.

4.7. In vivo Bioconversion of 3-(2,4-Dihydroxyphenyl)-propionic Acid

E. coli BL21 (DE3) bacteria, containing p4pmPmo and pTHPPDO plasmids, were grown in 200 mL of BHI medium at 30 °C and 180 rpm overnight. High density bacterial culture was centrifuged and resuspended in 200 mL of minimal C-750501 medium, in which synthesis of proteins was induced with 1 mM of IPTG at 20 °C and 180 rpm [65]. After 24 h of induction 3-(2,4-dihydroxyphenyl)-propionic acid was added to the final concentration of 4 mM. Bioconversion mixture was incubated for another 3 days at 30 °C with shaking. Cells were removed by centrifugation for 30 min at 3220× g and the supernatants were kept at 4 °C.

4.8. Purification of 3-(2,4-Dihydroxyphenyl)-propionic Acid Bioconversion Product

The supernatant containing the 3-(2,4-dihydroxyphenyl)-propionic acid oxidation product was incubated with 1.2 M ammonium chloride at room temperature overnight [39,50]. Reaction mixture was concentrated to ~100 mL volume and adjusted to pH 1 with concentrated HCl. The remains of substrate were extracted by five consecutive changes of 25 mL of ethyl acetate, and then aqueous fraction was purified using a reverse phase C_{18} column, equilibrated with water. Column was washed with at least 100 mL of water and then eluted with linear gradient of 0–60% methanol solution at a flow rate of 2 mL/min. Aqueous fractions were combined and evaporated (40 °C). Brownish crystals were dissolved in 0.1% formic acid solution and again loaded onto a reverse phase C_{18} column, previously equilibrated with 0.1% formic acid solution. Column was washed with at least 30 ml of 0.1% formic acid solution and then eluted with 60% methanol solution. Picolinic acid derivative was eluted with 0.1% formic acid solution. Fractions containing the product were collected, combined, and evaporated (40 °C). Picolinic acid yield from 145 mg of 3-(2,4-dihydroxyphenyl)-propionic acid fermentation was 34 mg, 24% of the theoretical yield. The product had traces of formic acid impurities, which aided the dissolution of the analyte in D_2O for NMR analysis.

4.9. High-Performance Liquid Chromatography and Mass Spectrometry

Before the analysis, the samples were mixed with an equal part of acetonitrile and centrifuged. High-performance liquid chromatography and mass spectrometry (HPLC-MS) was carried out using the system, consisting of the CBM-20 control unit, two LC-2020AD pumps, SIL-30AC auto sampler, and CTO-20AC column thermostat, using the SPD-M20A detector and LCMS-2020 mass spectrometer with ESI source (Shimadzu, Kyoto, Japan).

Chromatographic fractionation was conducted using YMC-Pack Pro C_{18} column, 150 × 3 mm (YMC, Kyoto, Japan) at 40 °C, with 0.1% formic acid solution in water and acetonitrile gradient from 5% to 95%.

Mass spectra were recorded from m/z 10 up to 500 m/z at 350 °C and ±4500 V using N_2. Mass spectrometry analysis was carried out using both the positive and negative ionization modes. The data were analyzed using LabSolutions LC/MS software (Shimadzu, Japan).

4.10. Nuclear Magnetic Resonance Spectroscopy

In total, 20 mg of the sample was dissolved in 0.5 mL D_2O. NMR spectra were recorded on an Ascend 400: ^1H NMR – 400 MHz, ^{13}C NMR – 100 MHz (Bruker, Billerica, MA, USA). Chemical shifts are reported in parts per million relative to the solvent resonance signal as an internal standard.

4.11. Protein MS-MS Analysis

Peptides were subjected to de novo sequencing based on matrix-assisted laser desorption ionization time of flight (MALDI-TOF/TOF) mass spectrometry (MS) and subsequent computational analysis at the Proteomics Centre of the Institute of Biochemistry, Vilnius University (Vilnius, Lithuania). The sample was purified as described previously in Materials and Methods. Tryptic digests (0.5 µL) were transferred on 384-well MALDI plate with 0.5 µL 4 mg/mL α-cyano-4-hydroxycinnamic acid (CHCA) matrix in 50% acetonitrile with 0.1% trifluoroacetic acid and analyzed with an Applied Biosystems/MDS SCIEX 4800 MALDI TOF/TOF™ mass spectrometer. Spectra were acquired in the positive reflector mode between 800 and 4000 m/z with fixed laser intensity at 3700 (Laser shots: 400, Mass accuracy: ±50 ppm). The most intense peaks of each survey scan (MS) were fragmented for sequence analysis (Collision energy: 1 keV; CID: no CID or medium air pressure CID used; Laser intensity: 4200–4400; Laser shots: 500–1000; Fragment mass accuracy: ±0.1 Da). Sequence analysis and peak lists were generated using GPS Explorer™ De Novo Explorer.

4.12. DNA Sequencing and Accession Numbers

The genome sequencing by Illumina (paired-ends) and assembling (in total, 35 contigs) was carried at Baseclear (Leiden, The Netherlands). The FASTQ sequence reads were generated using the Illumina Casava pipeline version 1.8.3. Initial quality assessment was based on data passing the Illumina Chastity filtering. Subsequently, reads containing adapters and/or PhiX control signal were removed using an in-house filtering protocol. The second quality assessment was based on the remaining reads using the FASTQC quality control tool version 0.10.0. The number of reads was 5,605,132 and an average quality score (Phred) was 37.72.

The accession number for partial 16S ribosomal RNA nucleotide sequence of *Pseudomonas* sp. 7HK4 is MH346031. Accession numbers for the sequences of *hcdA*, *hcdB*, and *hcdC* genes are MH346032, MH346033, MH346034, respectively.

Supplementary Materials: The following are available online, Table S1: Materials and reagents used in the studies, Table S2: Plasmids used in the studies, Table S3: The list of primers used in this study, Figure S1: Phylogenetic tree of *Pseudomonas* sp. 7HK4 bacteria based on partial 16S rDNA sequences, Figure S2: Biotransformation of 7-hydroxycoumarin, 6-hydroxycoumarin, coumarin and 6,7-dihydroxycoumarin by whole cells of *Pseudomonas* sp. 7HK4, Figure S3: Phylogenetic trees of HcdA, HcdB and HcdC proteins, Figure S4: SDS-PAGE, UV/Vis spectrum and analytical gel filtration chromatography of purified HcdA protein, Figure S5: Specificity of HcdA protein to flavin and nicotinamide cofactors, Figure S6: Activity of HcdA protein in different buffer systems, Figure S7: Activity of HcdA protein in different pH, Figure S8: Kinetic analysis of HcdA as determined by NADH oxidation for 3-(2,4-dihydroxyphenyl)-propionic acid, Figure S9: Kinetic analysis of HcdA as determined by NADH oxidation for NADH, Figure S10: Double reciprocal plot of NADH oxidation as a function of NADH concentration, Figure S11: Biotransformations of 3,4-dihydroxybenzoic acid, 2′,3′-dihydroxy-4′-methoxyaceto-phenone hydrate, gallacetophenone, pyrogallol, 2,3,4-trihydroxybenzoic acid and 2,3,4-trihydroxy-benzophenone by whole cells of *E. coli* BL21 containing *hcdB* gene, Figure S12: ^{1}H NMR spectrum of 6-(2-carboxyethyl)-4-oxo-1,4-dihydropyridine-2-carboxylic acid, Figure S13: ^{13}C NMR spectrum of 6-(2-carboxyethyl)-4-oxo-1,4-dihydropyridine-2-carboxylic acid, Figure S14: Resonance structure of *oxo*-picolinic acid derivative showing electron densities on aromatic carbons, Figure S15: SDS-PAGE of *E. coli* BL21 cell-free extract, containing induced recombinant HcdA, HcdB and HcdC proteins, Figure S16: HPLC-MS analysis of 3-(2-hydroxyphenyl)-propionic acid bioconversion mixture.

Author Contributions: Conceptualization and Funding Acquisition, R.M.; Investigation, A.K.; Data Analysis, A.K. and R.M.; Writing, Review & Editing, A.K. and R.M.

Funding: This research was funded by the European Union's Horizon 2020 research and innovation program [BlueGrowth: Unlocking the potential of Seas and Oceans] under grant agreement No. 634486 (project acronym INMARE).

Acknowledgments: We are grateful to Marija Ger for performing peptide sequence analysis, Daiva Tauraitė for help with chemical synthesis, and Laura Kalinienė for critical reading of the manuscript.

Conflicts of Interest: The authors declare no conflict of interest.

References

1. Sarker, S.D.; Nahar, L. Progress in the chemistry of naturally occurring coumarins. *Prog. Chem. Org. Nat. Prod.* **2017**, *106*, 241–304. [PubMed]
2. Murray, R.D.H.; Mendez, J.; Brown, S.A. *The Natural Coumarins—Occurrence, Chemistry and Biochemistry*; John Wiley: Chichester, UK, 1982.
3. Bourgaud, F.; Hehn, A.; Larbat, R.; Doerper, S.; Gontier, E.; Kellner, S.; Matern, U. Biosynthesis of coumarins in plants: A major pathway still to be unravelled for cytochrome P450 enzymes. *Phytochem. Rev.* **2006**, *5*, 293–308. [CrossRef]
4. Mazid, M.; Khan, T.A.; Mohammad, F. Role of secondary metabolites in defense mechanisms of plants. *Biol. Med.* **2011**, *3*, 232–249.
5. Kayser, O.; Kolodziej, H. Antibacterial activity of simple coumarins: Structural requirements for biological activity. *Z. Naturforsch. C* **1999**, *54*, 169–174. [CrossRef] [PubMed]

6. De Souza, S.M.; Delle Monache, F.; Smânia, A., Jr. Antibacterial activity of coumarins. *Z. Naturforsch C* **2005**, *60*, 693–700. [CrossRef] [PubMed]
7. Yang, L.; Ding, W.; Xu, Y.; Wu, D.; Li, S.; Chen, J.; Guo, B. New insights into the antibacterial activity of hydroxycoumarins against *Ralstonia solanacearum*. *Molecules* **2016**, *21*, 468. [CrossRef] [PubMed]
8. Serghini, K.; de Lugue, A.P.; Castejon, M.M.; Garcia, T.L.; Jorrin, J.V. Sunflower (*Helianthus annuus* L.) response to broomraoe (*Orobanche cernua* loefl.) parasitism: Induced synthesis and excretion of 7-hydroxylated simple coumarins. *J. Exp. Bot.* **2001**, *52*, 227–234. [CrossRef]
9. Shieh, H.S.; Blackwood, A.C. Use of coumarin by soil fungi. *Can. J. Microbiol.* **1969**, *15*, 647–648. [CrossRef] [PubMed]
10. Nakayama, Y.; Nonomura, S.; Tatsumi, C. The metabolism of coumarin by a strain of *Pseudomonas*. *Agric. Biol. Chem.* **1973**, *37*, 1423–1437. [CrossRef]
11. Levy, C.C.; Weinstein, G.D. The metabolism of coumarin by a microorganism. The reduction of o-coumaric acid to melilotic acid. *Biochemistry* **1964**, *3*, 1944–1947. [CrossRef] [PubMed]
12. Bellis, D.M. Metabolism of coumarin and related compounds in cultures of *Penicillium* species. *Nature* **1958**, *182*, 806–807. [CrossRef] [PubMed]
13. Sheila, M.B. The transformations of coumarin, o-coumaric acid and *trans*-cinnamic acid by *Aspergillus niger*. *Phytochemistry* **1967**, *6*, 127–130.
14. Aguirre-Pranzoni, C.; Orden, A.A.; Bisogno, F.R.; Ardanaz, C.E.; Tonn, C.E.; Kurina-Sanz, M. Coumarin metabolic routes in *Aspergillus* spp. *Fungal Biol.* **2011**, *115*, 245–252. [CrossRef] [PubMed]
15. Marumoto, S.; Miyazawa, M. Microbial reduction of coumarin, psoralen, and xanthyletin by *Glomerella cingulata*. *Tetrahedron* **2011**, *67*, 495–500. [CrossRef]
16. Levy, C.C.; Frost, P. The metabolism of coumarin by a microorganism. Melilotate hydroxylase. *J. Biol. Chem.* **1966**, *241*, 997–1003. [PubMed]
17. Roper, D.I.; Cooper, R.A. Purification, nucleotide sequence and some properties of a bifunctional isomerase/decarboxylase from the homoprotocatechuate degradative pathway of *Escherichia coli* C. *Eur. J. Biochem.* **1993**, *217*, 575–580. [CrossRef] [PubMed]
18. Díaz, E.; Timmis, K.N. Identification of functional residues in a 2-hydroxymuconic semialdehyde hydrolase. A new member of the alpha/beta hydrolase-fold family of enzymes which cleaves carbon-carbon bonds. *J. Biol. Chem.* **1995**, *270*, 6403–6411. [CrossRef] [PubMed]
19. Lim, J.C.; Lee, J.; Jang, J.D.; Lim, J.Y.; Min, K.R.; Kim, C.K.; Kim, Y. Characterization of the *pcbE* gene encoding 2-hydroxypenta-2,4-dienoate hydratase in *Pseudomonas* sp. DJ-12. *Arch. Pharm. Res.* **2000**, *23*, 187–195. [CrossRef] [PubMed]
20. Moonen, M.J.H.; Fraaije, M.W.; Rietjens, I.M.C.M.; Laane, C.; van Berkel, W.J.H. Flavoenzyme-catalyzed oxygenations and oxidations of phenolic compounds. *Adv. Synth. Catal.* **2002**, *344*, 1023–1035. [CrossRef]
21. Chaiyen, P. Flavoenzymes catalyzing oxidative aromatic ring-cleavage reactions. *Arch. Biochem. Biophys.* **2010**, *493*, 62–70. [CrossRef] [PubMed]
22. Crozier-Reabe, K.; Moran, G.R. Form Follows function: Structural and catalytic variation in the class A flavoprotein monooxygenases. *Int. J. Mol. Sci.* **2012**, *13*, 15601–15639. [CrossRef] [PubMed]
23. Romero, E.; Castellanos, J.R.G.; Gadda, G.; Fraaije, M.W.; Mattevi, A. Same substrate, many reactions: Oxygen activation in flavoenzymes. *Chem. Rev.* **2018**, *118*, 1742–1769. [CrossRef] [PubMed]
24. Abu-Omar, M.M.; Loaiza, A.; Hontzeas, N. Reaction mechanisms of mononuclear non-heme iron oxygenases. *Chem. Rev.* **2005**, *105*, 2227–2252. [CrossRef] [PubMed]
25. Cho, H.J.; Kim, K.; Sohn, S.Y.; Cho, H.Y.; Kim, K.J.; Kim, M.H.; Kim, D.; Kim, E.; Kang, B.S. Substrate binding mechanism of a type I extradiol dioxygenase. *J. Biol. Chem.* **2010**, *285*, 34643–34652. [CrossRef] [PubMed]
26. Van Berkel, W.J.; Kamerbeek, N.M.; Fraaije, M.W. Flavoprotein monooxygenases, a diverse class of oxidative biocatalysts. *J. Biotechnol.* **2006**, *124*, 670–689. [CrossRef] [PubMed]
27. Macheroux, P.; Kappes, B.; Ealick, S.E. Flavogenomics—A genomic and structural view of flavin-dependent proteins. *FEBS J.* **2011**, *278*, 2625–2634. [CrossRef] [PubMed]
28. Nakamura, S.; Nakamura, T.; Ogura, Y. Absorption spectrum of flavin mononucleotide semiquinone. *J. Biochem.* **1963**, *53*, 143–146. [CrossRef] [PubMed]
29. Koptyug, V.A. *Atlas of Spectra of Aromatic and Heterocyclic Compounds*; Science, Siberian Department of AS USSR: Novosibirsk, Russia, 1982.

30. Happe, B.; Eltis, L.D.; Poth, H.; Hedderich, R.; Timmis, K.N. Characterization of 2,2′,3-trihydroxybiphenyl dioxygenase, an extradiol dioxygenase from the dibenzofuran- and dibenzo-p-dioxin-degrading bacterium *Sphingomonas* sp. strain RW1. *J. Bacteriol.* **1993**, *175*, 7313–7320. [CrossRef] [PubMed]
31. Barnes, M.R.; Duetz, W.A.; Williams, P.A. A 3-(3-hydroxyphenyl) propionic acid catabolic pathway in *Rhodococcus globerulus* PWD1: Cloning and characterization of the *hpp* operon. *J. Bacteriol.* **1997**, *179*, 6145–6153. [CrossRef] [PubMed]
32. Vaillancourt, F.H.; Haro, M.A.; Drouin, N.M.; Karim, Z.; Maaroufi, H.; Eltis, L.D. Characterization of extradiol dioxygenases from a polychlorinated biphenyl-degrading strain that possess higher specificities for chlorinated metabolites. *J. Bacteriol.* **2003**, *185*, 1253–1260. [CrossRef] [PubMed]
33. Wolgel, S.A.; Dege, J.E.; Perkins-Olson, P.E.; Jaurez-Garcia, C.H.; Crawford, R.L.; Münck, E.; Lipscomb, J.D. Purification and characterization of protocatechuate 2,3-dioxygenase from *Bacillus macerans*: A new extradiol catecholic dioxygenase. *J. Bacteriol.* **1993**, *175*, 4414–4426. [CrossRef] [PubMed]
34. Asturias, J.A.; Eltis, L.D.; Prucha, M.; Timmisn, K.N. Analysis of three 2,3-dihydroxybiphenyl 1,2-dioxygenases found in *Rhodococcus globerulus* P6. *J. Biol. Chem.* **1994**, *269*, 7807–7815. [PubMed]
35. Bayly, R.C.; Dagley, S.; Gibson, D.T. The metabolism of cresols by species of *Pseudomonas*. *Biochem. J.* **1966**, *101*, 293–301. [CrossRef] [PubMed]
36. Burlingame, R.; Chapman, P.J. Catabolism of phenylpropionic acid and its 3-hydroxy derivative by *Escherichia coli*. *J. Bacteriol.* **1983**, *155*, 113–121. [PubMed]
37. Duggleby, C.J.; Williams, P.A. Purification and some properties of the 2-hydroxy-6-oxohepta-2,4-dienoate hydrolase (2-hydroxymuconic semialdehyde hydrolase) encoded by the Tol plasmid-pww0 from *Pseudomonas putida* mt-2. *J. Gen. Microbiol.* **1986**, *132*, 717–726. [CrossRef]
38. Riegert, U.; Heiss, G.; Fischer, P.; Stolz, A. Distal cleavage of 3-chlorocatechol by an extradiol dioxygenase to 3-chloro-2-hydroxymuconic semialdehyde. *J. Bacteriol.* **1998**, *180*, 2849–2853. [PubMed]
39. Wieser, M.; Eberspächer, J.; Vogler, B.; Lingens, F. Metabolism of 4-chlorophenol by *Azotobacter* sp. GP1: Structure of the *meta* cleavage product of 4-chlorocatechol. *FEMS Microbiol. Lett.* **1994**, *116*, 73–78. [CrossRef] [PubMed]
40. March, J. *Advanced Organic Chemistry Reactions, Mechanisms and Structure*, 3rd ed.; John Wiley & Sons Inc.: New York, NY, USA, 1985.
41. Gunther, H. *NMR Spectroscopy: Basic Principles, Concepts and Applications in Chemistry*, 3rd ed.; Wiley-VCH: Weinheim, Germany, 2013.
42. Simons, W.W. *The Sadtler Guide to Carbon-13 NMR Spectra*; Sadtler Research Laboratories: Philadelphia, PA, USA, 1983.
43. Retcofsky, H.L.; Friedel, R.A. *Carbon-13 Nuclear Magnetic Resonance Spectra of Monosubstituted Pyridines*; U.S. Dept of the Interior, Bureau of Mines: Washington, DC, USA, 1969.
44. Thomas, S.; Bruhl, I.; Heilmann, D.; Kleinpeter, E. 13C NMR chemical shift calculations for some substituted pyridines: A comparative consideration. *J. Chem. Inf. Comput. Sci.* **1997**, *37*, 726–730. [CrossRef]
45. Martin, G.; Dijols, S.; Capeillere-Blandin, C.; Artaud, I. Hydroxylation reaction catalyzed by the *Burkholderia cepacia* AC1100 bacterial strain. Involvement of the chlorophenol-4-monooxygenase. *Eur. J. Biochem.* **1999**, *261*, 533–539. [CrossRef] [PubMed]
46. Ricken, B.; Kolvenbach, B.A.; Corvini, P.F. *Ipso*-substitution—The hidden gate to xenobiotic degradation pathways. *Curr. Opin. Biotechnol.* **2015**, *33*, 220–227. [CrossRef] [PubMed]
47. Kolvenbach, B.A.; Corvini, P.F. The degradation of alkylphenols by *Sphingomonas* sp. strain TTNP3—A review on seven years of research. *N. Biotechnol.* **2012**, *30*, 88–95. [CrossRef] [PubMed]
48. Gabriel, F.L.; Heidlberger, A.; Rentsch, D.; Giger, W.; Guenther, K.; Kohler, H.P. A novel metabolic pathway for degradation of 4-nonylphenol environmental contaminants by *Sphingomonas xenophaga* Bayram: *Ipso*-hydroxylation and intramolecular rearrangement. *J. Biol. Chem.* **2005**, *280*, 15526–15533. [CrossRef] [PubMed]
49. Müller, R.; Lingens, F. Oxidative ring-cleavage of catechol in meta-position by superoxide. *Z. Naturforsch.* **1989**, *44c*, 207–211. [CrossRef]
50. Sala-Trepat, J.M.; Murray, K.; Williams, P.A. The metabolic divergence in the meta cleavage of catechols by *Pseudomonas putida* NCIB 10015. *Eur. J. Biochem.* **1972**, *28*, 347–356. [CrossRef] [PubMed]
51. Venugopala, K.N.; Rashmi, V.; Odhav, B. Review on natural coumarin lead compounds for their pharmacological activity. *Biomed. Res. Int.* **2013**, *13*, 14–19. [CrossRef] [PubMed]

52. Powell, J.A.; Archer, J.A. Molecular characterisation of a *Rhodococcus* ohp operon. *Antonie Van Leeuwenhoek* **1998**, *74*, 175–188. [CrossRef] [PubMed]
53. Díaz, E.; Ferrández, A.; Prieto, M.A.; García, J.L. Biodegradation of aromatic compounds by *Escherichia coli*. *Microbiol. Mol. Biol. Rev.* **2001**, *65*, 523–569. [CrossRef] [PubMed]
54. Arai, H.; Yamamoto, T.; Ohishi, T.; Shimizu, T.; Nakata, T.; Kudo, T. Genetic organization and characteristics of the 3-(3-hydroxyphenyl) propionic acid degradation pathway of *Comamonas testosteroni* TA441. *Microbiology* **1999**, *145*, 2813–2820. [CrossRef] [PubMed]
55. Müller, F.; Voordouw, G.; Van Berkel, W.J.H.; Steennis, P.J.; Visser, S.; Van Rooijen, P.J. A study of p-hydroxybenzoate hydroxylase from *Pseudomonas fluorescens*. *Eur. J. Biochem.* **1979**, *101*, 235–244. [CrossRef] [PubMed]
56. Shuman, B.; Dix, T.A. Cloning, nucleotide sequence, and expression of a p-hydroxybenzoate hydroxylase isozyme gene from *Pseudomonas fluorescens*. *J. Biol. Chem.* **1993**, *268*, 17057–17062. [PubMed]
57. Bunz, P.V.; Cook, A.M. Dibenzofuran 4,4a-dioxygenase from *Sphingomonas* sp. strain RW1: Angular dioxygenation by a threecomponent enzyme system. *J. Bacteriol.* **1993**, *175*, 6467–6475. [CrossRef] [PubMed]
58. Locher, H.H.; Leisinger, T.; Cook, A.M. 4-Sulphobenzoate 3,4-dioxygenase. Purification and properties of a desulphonative two-component enzyme system from *Comamonas testosteroni* T-2. *Biochem. J.* **1991**, *274*, 833–842. [CrossRef] [PubMed]
59. Vatsis, K.P.; Coon, M.J. Ipso-substitution by cytochrome P450 with conversion of p-hydroxybenzene derivatives to hydroquinone: Evidence for hydroperoxo-iron as the active oxygen species. *Arch. Biochem. Biophys.* **2002**, *397*, 119–129. [CrossRef] [PubMed]
60. Zhao, Z.; Koeplinger, K.A.; Peterson, T.; Conradi, R.A.; Burton, P.S.; Suarato, A.; Heinrikson, R.L.; Tomasselli, A.G. Mechanism, structure–activity studies, and potential applications of glutathione S-transferase-catalyzed cleavage of sulfonamides. *Drug Metab. Dispos.* **1999**, *27*, 992–998. [PubMed]
61. Corvini, P.F.X.; Meesters, R.J.W.; Schäffer, A.; Schröder, H.F.; Vinken, R.; Hollender, J. Degradation of a nonylphenol single isomer by *Sphingomonas* sp. strain TTNP3 leads to a hydroxylation-induced migration product. *Appl. Environ. Microbiol.* **2004**, *70*, 6897–6900. [CrossRef] [PubMed]
62. Arnold, R.T.; Buckley, J.S., Jr.; Richter, J. The dienone–phenol rearrangement. *J. Am. Chem. Soc* **1947**, *69*, 2322–2325. [CrossRef]
63. Spence, E.L.; Kawamukai, M.; Sanvoisin, J.; Braven, H.; Bugg, T.D. Catechol dioxygenases from *Escherichia coli* (MhpB) and *Alcaligenes eutrophus* (McpI): Sequence analysis and biochemical properties of a third family of extradiol dioxygenases. *J. Bacteriol.* **1996**, *178*, 5249–5256. [CrossRef] [PubMed]
64. Prieto, M.A.; Díaz, E.; García, J.L. Molecular characterization of the 4-hydroxyphenylacetate catabolic pathway of *Escherichia coli* W: Engineering a mobile aromatic degradative cluster. *J. Bacteriol.* **1996**, *178*, 111–120. [CrossRef] [PubMed]
65. Sivashanmugam, A.; Murray, V.; Cui, C.; Zhang, Y.; Wang, J.; Li, Q. Practical protocols for production of very high yields of recombinant proteins using *Escherichia coli*. *Protein Sci.* **2009**, *18*, 936–948. [CrossRef] [PubMed]
66. Laemmli, U.K. Cleavage of structural proteins during the assembly of the head of bacteriophage T4. *Nature* **1970**, *227*, 680–685. [CrossRef] [PubMed]
67. Lowry, O.H.; Rosebrough, N.J.; Farr, A.L.; Randall, R.J. Protein measurement with the folin-phenol reagents. *J. Biol. Chem.* **1951**, *193*, 265–275. [PubMed]
68. Gundry, R.L.; White, M.Y.; Murray, C.I.; Kane, L.A.; Fu, Q.; Stanley, B.A.; Van Eyk, J.E. Preparation of proteins and peptides for mass spectrometry analysis in a bottom-up proteomics workflow. *Curr. Protoc. Mol. Biol.* **2009**, *88*, 10.25.1–10.25.23.

Sample Availability: Samples of the plasmid DNAs are available from the authors.

© 2018 by the authors. Licensee MDPI, Basel, Switzerland. This article is an open access article distributed under the terms and conditions of the Creative Commons Attribution (CC BY) license (http://creativecommons.org/licenses/by/4.0/).

MDPI
St. Alban-Anlage 66
4052 Basel
Switzerland
Tel. +41 61 683 77 34
Fax +41 61 302 89 18
www.mdpi.com

Molecules Editorial Office
E-mail: molecules@mdpi.com
www.mdpi.com/journal/molecules

www.ingramcontent.com/pod-product-compliance
Lightning Source LLC
LaVergne TN
LVHW070612100526
838202LV00012B/629